REYKJAVIK AND BEYOND

Deep Reductions in Strategic Nuclear Arsenals and the Future Direction of Arms Control

Committee on International Security and Arms Control

National Academy of Sciences

NATIONAL ACADEMY PRESS
Washington, D.C. 1988

NATIONAL ACADEMY PRESS • 2101 Constitution Avenue, NW • Washington, DC 20418

NOTICE: The project that is the subject of this report was approved by the Governing Board of the National Research Council, whose members are drawn from the councils of the National Academy of Sciences, the National Academy of Engineering, and the Institute of Medicine. The members of the committee responsible for the report were chosen for their special competences and with regard for appropriate balance.

This report has been reviewed by a group other than the authors according to procedures approved by a Report Review Committee consisting of members of the National Academy of Sciences, the National Academy of Engineering, and the Institute of Medicine.

The National Academy of Sciences is a private, nonprofit, self-perpetuating society of distinguished scholars engaged in scientific and engineering research, dedicated to the furtherance of science and technology and to their use for the general welfare. Upon the authority of the charter granted to it by the Congress in 1863, the Academy has a mandate that requires it to advise the federal government on scientific and technical matters. Dr. Frank Press is president of the National Academy of Sciences.

The National Academy of Engineering was established in 1964, under the charter of the National Academy of Sciences, as a parallel organization of outstanding engineers. It is autonomous in its administration and in the selection of its members, sharing with the National Academy of Sciences the responsibility for advising the federal government. The National Academy of Engineering also sponsors engineering programs aimed at meeting national needs, encourages education and research, and recognizes the superior achievements of engineers. Dr. Robert M. White is president of the National Academy of Engineering.

The Institute of Medicine was established in 1970 by the National Academy of Sciences to secure the services of eminent members of appropriate professions in the examination of policy matters pertaining to the health of the public. The Institute acts under the responsibility given to the National Academy of Sciences by its congressional charter to be an adviser to the federal government and, upon its own initiative, to identify issues of medical care, research, and education. Dr. Samuel O. Thier is president of the Institute of Medicine.

The National Research Council was organized by the National Academy of Sciences in 1916 to associate the broad community of science and technology with the Academy's purposes of furthering knowledge and advising the federal government. Functioning in accordance with general policies determined by the Academy, the Council has become the principal operating agency of both the National Academy of Sciences and the National Academy of Engineering in providing services to the government, the public, and the scientific and engineering communities. The Council is administered jointly by both Academies and the Institute of Medicine. Dr. Frank Press and Dr. Robert M. White are chairman and vice chairman, respectively, of the National Research Council.

This work was supported by a grant to the National Academy of Sciences from the John D. and Catherine T. MacArthur Foundation.

Library of Congress Catalogue Card No. 87-30193

ISBN 0-309-03799-9

Printed in the United States of America

Seminar Participants

THE HONORABLE FRED C. IKLÉ,* Under Secretary of Defense for Policy

MARVIN L. GOLDBERGER,† Director, Institute for Advanced Study; Former Chairman, Committee on International Security and Arms Control

COMMITTEE ON INTERNATIONAL SECURITY AND ARMS CONTROL

WOLFGANG K. H. PANOFSKY† (*Chairman*), Director Emeritus, Stanford Linear Accelerator Center, Stanford University

LEW ALLEN, JR., Director, Jet Propulsion Laboratory, California Institute of Technology

PAUL M. DOTY,† Director Emeritus, Center for Science and International Affairs, Harvard University

ALEXANDER H. FLAX,† President Emeritus, Institute for Defense Analyses

EDWARD A. FRIEMAN, Director, Scripps Institution of Oceanography

RICHARD L. GARWIN, Science Advisor to the Director of Research, Thomas J. Watson Research Center, IBM Corporation

SPURGEON M. KEENY, JR.,† President, Arms Control Association

CATHERINE M. KELLEHER,† Director, Maryland International Security Project, University of Maryland

JOSHUA LEDERBERG, President, Rockefeller University

CLAIRE MAX, Associate Director, Institute of Geophysics and Planetary Physics, Lawrence Livermore Laboratory

MICHAEL M. MAY, Associate Director at Large, Lawrence Livermore Laboratory

RICHARD A. MULLER, Lawrence Berkeley Laboratory, University of California

JOHN D. STEINBRUNER,† Director, Foreign Policy Studies Program, Brookings Institution

---

* Seminar speaker; declined to have his talk included in this publication.

† Seminar speaker.

Foreword

Since its creation in 1864, the National Academy of Sciences has undertaken many studies and activities relating to matters of national security, and currently several committees of the National Research Council advise branches of the military on questions of scientific research. Other Academy committees have studied topics such as nuclear winter and the contribution of behavioral and social sciences to the prevention of nuclear war.

The Committee on International Security and Arms Control (CISAC) reflects the Academy's deep interest in international security and the potential of arms control to reduce the threat of nuclear war. Its members have been deeply involved in many aspects of military technology and arms control. They have advised several presidents and served in senior governmental posts; they have been involved in military research since the days of the Manhattan Project; they have headed universities and research centers; they have been involved with important arms control negotiations. The members of this committee have thought long and hard about national security issues.

The committee has pursued a number of activities in response to its broad charter. Twice each year it meets with its counterparts from the Soviet Academy of Sciences to explore problems of international security and arms control. In response to the widely expressed interest of Academy members in learning more about issues and opportunities in arms control, it has convened a number of meetings and sessions on arms control specifically for them. In the spring of 1984 CISAC conducted a major tutorial for over 200 Academy members. The background materials for that tutorial resulted in the book *Nuclear Arms Control: Background and Issues,* published in 1985. CISAC also conducted a seminar on strategic defense in 1985 and cosponsored one the following year on crisis management that resulted in the short publication *Crisis Management in the Nuclear Age.*

In the spring of 1987 CISAC presented a seminar for the Academy audience that explored the implications of the proposals for very deep cuts in strategic nuclear arsenals that had been discussed by President Reagan and General Secretary Gorbachev at the Reykjavik summit in 1986. The committee felt that, whereas many people instinctively support the goal of significantly reducing arsenals, very little serious study had been done on what that would actually mean and on how very deep cuts would affect other aspects of the military balance and the political and international order more broadly. CISAC members thus shared their initial thoughts on what changes in force structures, strategic thought, and political relations would be necessary to make possible large reductions in the superpowers' nuclear arsenals.

Because the response to this seminar was so positive, I asked that the talks be collected in a small volume that could be shared with a wider audience. I believe this volume provides a useful starting point for thinking about how to tackle the difficult political and military issues that arise in contemplating the transition to a safer world with significantly fewer nuclear weapons—a goal that has been enunciated by both the American and Soviet leaders and embraced by citizens everywhere.

I would like to express my great appreciation to the chairman and members of CISAC, some of whom contributed to this volume and all of whom dedicate much time and effort to the activities of the committee. I believe the committee continues to learn a great deal in the course of its work, and I hope that others will judge that work, including this volume, to be useful in their own effort to understand the role of arms control in reducing the threat of nuclear war.

FRANK PRESS, *President*
National Academy of Sciences

Contents

REYKJAVIK AND BEYOND

1

Reykjavik and Beyond:
Implications of Deep Reductions in Strategic Nuclear Arsenals and the Future Direction of Arms Control

Wolfgang K. H. Panofsky

On March 23, 1983, President Ronald Reagan appealed to scientists to make "nuclear weapons impotent and obsolete." He said: "Is it not worth every investment necessary to free the world from the threat of nuclear war?" On January 15, 1986, General Secretary Mikhail Gorbachev made a worldwide appeal for the abolishment of nuclear weapons. Both leaders based their statements on a common conviction that the doctrine of nuclear deterrence could not be a permanent basis for our security. Recently, during the large meeting of dignitaries and scientists on February 14, 1987, Gorbachev stated: "There is probably no one in this hall or elsewhere who considers nuclear weapons innocuous; however, quite a few people sincerely believe them an evil necessary to prevent a greater evil—war." He then continued: "We would have to admit that the nuclear safeguard is not fail-safe or of endless duration."

Thus, the two leaders appear to agree on the ends in this matter, but they drastically diverge on the means. The Committee on International Security and Arms Control (CISAC) of the National Academy of Sciences devoted a seminar at the Academy's 1985 annual meeting to a discussion of whether dedicated strategic defenses deployed against ballistic missiles could meet President Reagan's objective. The general conclusion of that seminar and the emerging consensus of almost all informed individuals in the technical community is that unilaterally deployed strategic defense does not constitute either a feasible or stable path toward the elimination of nuclear weapons. However, there remains some disagreement as to what role, if any, strategic defenses should play when we consider lesser objectives, or in the event that they were deployed by mutual consent after large reductions on both sides.

At Reykjavik the two leaders reached consensus again on the goal of eliminating some categories of nuclear weapons. There appeared to be disagreement on which categories were to be eliminated and there continued to be major disagreement on strategic defense. After the meeting the U.S. position was that in 10 years strategic ballistic missiles should be eliminated; the Soviet position called for the elimination of all strategic nuclear weapons. What was actually agreed to during the meeting on this subject remains to some extent contentious and at this point possibly not relevant.

What this dialogue has done, however, is to stimulate more intensive deliberation within the military establishment, among the United States and its allies, and within that part of the intellectual community concerned with strategic matters on which paths might be followed toward drastic reductions of nuclear weapons systems. It is somewhat paradoxical that, whereas the arms control community has dedicated considerable effort to analyzing and generally criticizing strategic defense, little has been done to study the more customary path to arms control; that is, limiting, decreasing, and eventually eliminating certain categories of weapons. Analysts have stayed clear of really facing the possibility of major success along this road; in other words, the community has considered the possibility of making nuclear weapons "impotent and obsolete" through defenses and most have found the concept wanting, but it has not analyzed the possibility of achieving success along the direction of the traditional path of arms control.

This seminar is dedicated to exploring that latter direction as triggered by the deliberations at Reykjavik. Let me warn you from the outset not to expect any ringing declarations or the identification of clear solutions. Although the problem is to some extent tractable to a level of arms reductions that goes much further than those now on the bargaining table in Geneva, it is difficult to foresee the total elimination of nuclear weapons from this earth without drastic changes in the international order.

The reason for this somewhat pessimistic conclusion lies in the nature of nuclear weapons. It is not based on any particular strategic or political doctrine or policy; neither does it depend on any particular form of social organization as long as sovereign states remain. Nuclear weapons have increased by a factor of over one million the amount of explosive power that can be concentrated into a weapon of a given size and weight. Thus, the delivery and explosion of even a quite small number of nuclear weapons can wreak unspeakable havoc. As discussed in previous seminars, this has not only greatly tilted the traditional balance between offense and defense in favor of the offense, but it has also made the power of even a tiny fraction of the world's existing offensive arsenals extremely large. These two factors have the consequence that, however distasteful or even immoral the so-called "doctrine" of nuclear deterrence may be to some, one has to conclude that, considering the size of today's nuclear stockpiles, it is not really a doctrine but a physical fact beyond the reach of political authority to deny. Therefore, the context of the following discussion is that mutual deterrence will be a fact of life for the foreseeable future but that a great deal can be done to decrease both the burdens and dangers of nuclear armament and nuclear war.

Last year's annual meeting seminar dealt with one facet of the problem that is at some level independent of the nature of the arsenals themselves: crisis management in the nuclear age. Today's seminar examines the path toward deep reductions of nuclear weapons.

Although the Reykjavik Summit has provided the incentive for analyzing this question in greater depth, the meeting had a mixed outcome. It broke up in discord on the matter of strategic defense, and there were no documented specific agreements. Yet both leaders, for political reasons of their own, declared the meeting a success after a brief period of apparent dejection. Again, in the words of Gorbachev during the February 17, 1987, gathering of luminaries,

"It was not a success, it was a breakthrough . . . a momentous opportunity to embark on the path leading to a nuclear weapon-free world was glimpsed." Statements within the West were less monolithic. Some Defense Department spokesmen attacked the "nuclear-free" concept outright and European leaders suddenly found themselves confronted by a possible folding up of the nuclear umbrella, which would leave them with a more uncertain future determined by the balance of conventional forces.

Let me review briefly the actual accomplishments and nonaccomplishments from Reykjavik, using what I might call a "score sheet" of the deliberations.

As shown in Figure 1, the results from Reykjavik can be categorized into three essential segments:

1. Agreement on limiting or eliminating intermediate nuclear forces in Europe and reducing central strategic systems by what is billed to be 50 percent but that is actually somewhat less. This program is to be accomplished fully in 5 years.
2. To reduce "something" to zero by the end of a 10-year period, whether that something is ballistic missiles, according to the U.S. version, or strategic nuclear weapons, according to the Soviet version.
3. Proposals on strategic defenses from the two sides, resulting in unresolved disagreement at the end of the meeting.

The first set of agreements is generally billed as a success, although one should recall that a great deal remained undiscussed or only partly discussed at Reykjavik. There was agreement on basic numbers. In particular, the Soviets accepted what was originally the U.S. "zero-zero" proposal for zero intermediate-range nuclear forces (INF) deployment in Europe, with each side retaining 100 warheads that could be deployed within the United States and East Asia, respectively. (Both sides have since agreed not to retain the 100 INF missiles outside of Europe.) Progress was made on agreements for the reduction of central strategic systems, in particular with respect to the hitherto controversial issue of how to count nuclear weapons on strategic bombers. What was left were verification issues, sea-launched cruise missiles, and the incorporation of the fate of shorter-range missiles in the INF agreement. At the time of the Reykjavik meeting the Soviets insisted that progress along all these lines should

	REPORTED "AGREEMENT"	OPEN ISSUES
Intermediate-Range Nuclear Forces (INF)	No INF delivery vehicles in Europe; 100 warheads permitted elsewhere	Verification details Shorter-range missiles
Strategic Arms Reductions Talks (START) In 5 years	1,600 Strategic delivery vehicles (cut from 2,400) 6,000 Warheads (cut from 10,000) Counting rules for weapons on strategic bombers	Subcategories Verification Submarine-launched cruise missiles
In 10 years	Zero strategic weapons or zero ballistic missiles Aircraft, cruise missiles not specified	How to get there? Verification? Conventional balance?
Comprehensive Test Ban (CTB)	Start talking again	No CTB until ballistic missiles eliminated (United States) Staging? Verification?

	U.S. POSITION	SOVIET POSITION
Strategic Defense Initiative (SDI)	"New" interpretation of ABM treaty (permits testing and development) Abrogate ABM treaty in 10 years; deploy ABM system if developed	Laboratory work only, and (?) traditional interpretation of ABM treaty, such as testing from fixed sites Discuss future of ABM treaty in 10 years; treaty still in force

FIGURE 1 The Reykjavik "score sheet."

be linked to an agreement on ballistic missile defense, and shortly after Reykjavik they deepened their commitment to that linkage. Only the subsequent agreement by the Soviets to unlink consideration of INF from limits on the Strategic Defense Initiative (SDI) has led to the recent discussions that are viewed with so much hope by the world and may result in a signed INF treaty by the end of 1987.

The SDI discussions were totally unproductive. I consider the U.S. position to be essentially a move toward destroying the antiballistic missile (ABM) treaty. We agreed with the Soviets to abide by the treaty for 10 years; that is, not to invoke for a decade the abrogation provisions spelled out in it. However, although this was not explicitly stated, we asserted the right to live within the treaty strictures using the so-called broad interpretation, which permits unlimited testing and development of weapons in space. Moreover, according to our position, the ABM treaty would be definitely abrogated after 10 years, whereas according to the Soviet position the continuation of the treaty after 10 years would be a subject of discussion.

Because as a practical matter deployment of space-based ABM weapons realistically could not be accomplished for 10 years at any rate, the practical consequence of the U.S. position, had it been accepted, would have been an immediate abrogation of the ABM treaty. The Soviet position has been reported as one of wishing to strengthen the ABM treaty by restricting research and development work to the laboratory only. There remains some ambiguity as to whether the word "strengthen" used by the Soviets really meant to modify the treaty or to support it firmly in its traditional form. There is also ambiguity as to whether the word "laboratory" is meant to restrict experimentation to the strictly "under roof" type or whether it would include, for instance, "laboratories" in space or some other, more generic interpretation. Trying to define how the provisions of the ABM treaty in its traditional interpretation apply specifically to various technologies remains a major challenge to the U.S.-USSR dialogue. Introducing the broad interpretation, or what the United States euphemistically calls the "legally correct" interpretation, would destroy the assumptions on which the U.S. Senate based its ratification process and contradicts the statements and testimony of the U.S. participants in the negotiations and the understanding that has been maintained since 1972.

The discussions on the total elimination of "something" nuclear bear witness to the parties' lack of preparation for the summit. Clearly, this was neither the time nor the place to reach an agreement on something as dramatic as zero nuclear ballistic missiles, or even zero strategic nuclear delivery systems, without prior consultation with the Allies, the Joint Chiefs of Staff, and Congress. Not surprisingly, the resultant agreement has drawn protests from all of these constituencies, and spokesmen for the White House have stated that they would "deemphasize" that subject in future talks with the Soviets.

Nevertheless, something very positive was achieved as a result of these zero discussions: public dialogue of quite drastic reductions has been made respectable. Clearly, the objections voiced by the Allies, Congress, the academic strategic community, and even spokesmen within the administration have a sound basis. Yet these objections should be discussed and examined on their merits. There are indeed many questions associated with deep reductions of central strategic forces. For example, there is the resulting increased role of the remaining shorter-range nuclear weapons in Europe. There is the question of verification; that is, whether or not relatively small clandestine deployments would have great leverage. There is the question of whether increasing the role of strategic defenses might destabilize the situation if the offensive balance is sustained at lower force levels. There is the issue of the conventional force balance in Europe, which would come more into the forefront if the nuclear umbrella—the last recourse in case of threatened defeat—is no longer in place. There is the problem of the nuclear forces of the Allies and China, which would become more prominent as the superpowers' arsenals were reduced.

Not only do these questions come to the forefront as central strategic nuclear forces are reduced, but the basic concepts of deterrence also require reexamination. This should not be viewed as a problem but rather as a challenge—such a reexamination is long overdue. Our basic concepts of deterrence have drifted over the last decades. Originally one would consider an opponent to be deterred if by attacking first, it would face "unacceptable damage" to its society through retaliation by the surviving forces of its adversary. Doctrines then shifted to the "flexible response," which implied that retaliation would not require an all-or-nothing strike but could be tailored in magnitude to the situation faced. Flexible response gave way to "selective targeting," which meant deterrence by threatening to use a small number of sorties that would permit very selective attacks on fixed enemy military or industrial targets. The concept has changed still further to the requirement that deterrence deny the opponent, after it attacked, the opportunity to continue the war; in other words, a second strike would have to be strong enough to destroy the opponent's war-making potential. This progression of doctrines requires translation into actual military planning; in regard to the potential uses of strategic nuclear forces, such a translation means the establishment of target lists. The target lists in turn are translated into requirements for strategic nuclear strike delivery systems in a somewhat arbitrary process. One of the fundamental reasons why we have been unable to answer the question "When is enough enough?" to the satisfaction of the military constituencies is the evolution of such doctrines, all in the name of deterrence.

Deterrence is a state of mind and is not based on specifiable physical facts. It assumes that a rational decisionmaker would not be tempted to initiate a nuclear war in time of crisis or if he or she sees an evolving uncontrollable threat. Yet there can be and is a wide range of judgment as to what it would take to deter the opponent under such circumstances, even if you assume that the leaders of the opposing nation are rational. If you go beyond that and assume that the opponent's leaders are not rational, then the concept of deterrence becomes totally indefinable.

The levels of forces required for deterrence vary drastically depending on which of the abovementioned interpretations of deterrence you accept. Under the earlier definitions of deterrence, it was argued, and I believe reasonably, that the potential of 30 million–50 million dead would be adequate deterrence; yet the nuclear weapons unleashed from only a single Trident submarine against major Soviet cities would have just that effect. Flexible response leads to somewhat larger force requirements but mainly challenges command, control, and decision-making bodies. Denying the ability to continue the war is fundamentally an open-ended prescription for additional requirements.

Few words in the military jargon have been abused as much as the term "requirement." One hears, for instance, such comments as, "If we reduce our forces from the present 10,000-plus strategic nuclear weapons to the 6,000 weapons agreed to at Reykjavik, then only 75 percent of our 'required' targets can be struck." Anyone familiar with the targeting situation knows that lists include many potential targets of progressively decreasing military significance. In a real way there continues to be a symbiotic relationship between requirements and the available weapons systems—military targeting requirements lead to the acquisition of new weapons, and the existence of new military delivery systems leads to a search for new targets. Thus, one of the major benefits of facing the prospect of drastic reductions would be an examination of just such targeting priorities and strategic doctrines. If there were fewer weapons available for deterrent purposes, fewer weapons could be dedicated to missions and targets that are superfluous and in fact dangerously destabilizing to the fundamental concept of deterring the opponent from initiating nuclear war.

In CISAC's deliberations on this question, we have carried out numerical studies on purely countermilitary exchanges to examine the retaliatory potential remaining if the current number of strategic weapons was reduced to the 6,000 warheads agreed to at Reykjavik or by another factor of 2, which might be an example of a more drastic regimen of cuts. Quite apart from the other issues already cited, this lower level is likely to deny coverage of some military targets now included in a scenario intended to deny the opponent the opportunity to continue fighting the war. Yet when one examines, even at the 3,000-warhead level, the collateral civilian deaths that would result in this type of intended, purely antimilitary exchange, one obtains numbers in the many tens of millions. Thus, one may legitimately ask: "Is the decision faced by the national leadership whether and how to retaliate against nuclear attack really that different whether one lives within an 'unacceptable damage,' 'flexible response,' 'selective targeting,' or 'denial of continued war-fighting' doctrine?" Even more important may be whether the willingness of national leaders to initiate nuclear war in the face of certain retaliation is really that dependent on the perceived doctrine that the opponent might follow. In short, have we not really "doctrined" ourselves into progressively increasing military requirements for deterrence without examining whether this makes military sense, quite apart from the basic inhumanity of this line of thinking? A reexamination of these questions is essential if meaningful discussions on drastic cuts are to flow from the Reykjavik experience.

One should now dare to inquire what changes in strategic thought and even in the international order are required to make drastic cuts acceptable. This seminar is dedicated to the examination of the totality of these issues. Let me say again that the audience should have no illusion that the speakers have answers to these many questions or that a clear message will emerge from this discussion. However, I sincerely hope that this seminar will play a significant role in adding an independent, reasonable voice to supplement the chorus demanding a world moving toward freedom from nuclear weapons.

2

The Purpose and Effect of Deep Strategic Force Reductions

John D. Steinbruner

My topic is a discussion of the calculations of the National Academy of Sciences' Committee on International Security and Arms Control regarding the implications of strategic force reductions. I will emphasize the results of the calculations rather than their internal details.

In assessing the effect of reducing force balances, the first step, of course, is to determine the purposes that are to be achieved. We made the judgment that the primary objective in seeking force reductions should be that of constraining the capacity, and therefore the inclination, to undertake an effective preemptive attack on the opponent's strategic forces. Although it has not been as explicitly defined as one might wish, we believe there is an implicit consensus in support of this objective.

The capacity for preemption is not considered a legitimate security requirement under current international understanding. However they might grumble about it, both the United States and the Soviet Union in fact concede to each other the legitimacy of having a deterrent force that promises an effective retaliation to an attack, but they certainly deny the legitimacy of having a preemptive attack capability that could remove the deterrent requirement.

If there is a level of forces that would guarantee the ability to perform deterrent missions but preclude the capacity to threaten the opponent's deterrent force, presumably that would be the basis for stable agreement. Our calculations attempt to define such a force level.

To do that, one must judge the deterrent requirement. Dr. Wolfgang Panofsky provided an outline of the logic that has been used in the past to define this requirement. Let me focus on the high and low ends of prevailing opinion in that regard and try to give a loose quantitative definition of the range, while recognizing that there are some who would put the requirements for deterrence outside of these low and high ends of what I believe to be the prevailing mainstream opinions on this matter.

The high side—that is, the more demanding notion of what deterrence requires—is determined by the prevailing doctrine inside the military organizations that actually target nuclear weapons. Their idea is that effective retaliation must attack the opposing military infrastructure and essentially remove its capabilities. That is a fairly demanding requirement in terms of the number of weapons and the locations in which they must be put.

On the low side of the current range of opinion, there is the notion that a capacity to destroy the social infrastructure of the opponent would be enough to deter war. The implicit requirement in this case is the ability to attack urban-industrial areas, causing tens of millions of deaths and destroying some 75 percent of basic industrial facilities.

In order to associate rough numbers with these theories, our calculations estimated the array of military and economic targets that actually exist in the United States and the Soviet Union. We considered 6,000 of those targets because our initial purpose was to consider, roughly, the effects of reducing forces to 6,000 warheads. We adopted the simpleminded notion that each of those weapons should have a target associated with it, and we imagined the targets that would be assigned by a Soviet or an American military commander.

These estimates reveal the phenomenon that Dr. Panofsky mentioned. Beyond a certain level of forces—and it falls well below 6,000—the targets get smaller and smaller and less and less significant in terms of their contribution to the cumulative level of damage. This is the economist's familiar notion of diminishing margin of utility.

With that effect in mind, we defined the idea of an efficient deterrent requirement. An efficient deterrent requirement is limited by the diminishing significance of attacking additional targets after a certain number have been destroyed. Beyond that point a prudent military commander should prefer to save those weapons rather than use them because the marginal damage they would inflict would not justify expending the weapons.

We tried to define that number using the more demanding criterion of retaliation against the residual military capability of the opponent. We examined the infrastructure of conventional and strategic forces and military-support industries in the United States and the Soviet Union and determined that beyond the first 2,000–3,000 targets, additional attacks would be of so little marginal significance that a prudent commander would not continue the attack beyond those levels. So much of the military and social infrastructure would have been destroyed that additional destruction would not have sufficient military effect to justify the expenditure of the weapons.

We assessed the effect of an attack on the approximately 2,000 highest priority military and economic infrastructure targets using standard calculations to estimate civilian fatalities—not casualties but rather immediate, prompt deaths from such an attack. We found that a successful retaliatory attack on these 2,000 military and economic targets would kill 20 million–40 million people in prompt blast damage effects in the United States and 30 million–50 million in the Soviet Union. As it turns out, those same ranges for prompt civilian fatalities appear if one assesses what would happen if the attack were directed to urban-industrial areas using a much smaller number of weapons. An attack with a 100-equivalent-megaton total yield delivered by a few hundred weapons directed against the urban-industrial areas of the United States or the Soviet Union would produce prompt fatalities from blast and thermal pulse effects in these ranges as well.

Therefore, as Dr. Panofsky pointed out, the two ends of the

spectrum of opinion about deterrent requirements do agree on that number. If retaliation is directed against urban-industrial areas, according to the minimum deterrence theory, or if retaliation is directed against the larger target set required to destroy the residual military power of the enemy, a total of 50 million–90 million people will die in the two countries; and, of course, the casualties and social disruption would be some multiple of that figure.

In sum, then, 500–2,000 warheads delivered in retaliation covers anything that might be considered a reasonable deterrent requirement under any of the prevailing opinions about that requirement.

With that thought, we went on to assess how forces might be reduced in a way that allowed this core deterrent requirement to be maintained by both sides while eliminating the capacity of either one to attack the forces that would conduct this retaliation. We examined seven ways in which current U.S. and Soviet forces could be reduced to the level of 6,000 warheads. A number of very simple observations emerged from that exercise.

First, no matter how forces might be reduced to the 6,000-warhead level, each side can cover the estimated efficient deterrent requirements either in retaliation or under a launch-on-warning strategy. Even in the worst case, in which the victim of an initial attack waits until the attack is completed before beginning to retaliate, a 6,000-warhead force would still be able to cover these requirements. There is not much pressure, then, on the judgments made in reducing forces to the 6,000-warhead level. However it is done, both sides can be confident that they can meet deterrent requirements.

The main effect of different patterns of reductions concerns the residual forces that would remain after both sides had completed their initial attacks. The range of models we considered includes one in which we preserved all the multiple-warhead, high-accuracy systems on both sides, thereby maximizing preemptive capability. According to the central criterion we set to guide the analysis, that pattern of reduction would be the most perverse case.

Even under that case, both sides can meet their deterrent requirements, but there would be some theoretical incentive for preemption in the sense that residual forces, after meeting those requirements, would not be even. The side that went first would have more weapons left after the exchange than the side that went second, and military organizations might care about such a thing, although it is doubtful that the public would.

Because military organizations are likely to be the groups responsible for assessing the possible patterns of force reductions, the reductions that are more likely to be implemented involve the removal of those weapons that are most easily or most likely to be used for a preemptive attack: notably, the SS-18 on the Soviet side and the MX, and potentially the Trident 2 missile, on our side.

If the weapons were eliminated in reducing strategic forces to the 6,000-warhead level, not only would both sides be able to meet efficient deterrent requirements but they would also expect to have equal forces after the initial exchange. Under this arrangement it would not matter who goes first and who goes second; both sides would in the end be about the same, which is very badly off indeed.

We next tested the effect of reducing strategic forces another 50 percent—to the level of 3,000 warheads on each side. We wanted to determine whether the capacity to cover the full range of deterrent targets at these yet lower force levels would depend on questionable values for the operational assumptions used in the calculations. We judged that, if the robustness of the result did depend on input assumptions that were subject to plausible disagreement, it would be quite difficult for force reductions of this magnitude to achieve the political support necessary to carry them out. The operational assumptions in question concerned alert rates for bombers and submarines, the time required for bombers to fly out of the zone of vulnerability around their base areas, and the warning they would enjoy in doing so, as well as the standard characteristics of weapons—their reliability, yield, and accuracy. We also assumed that all the forces of the attacker would be fully prepared whereas the victim would have to respond with forces available under daily conditions. That assumption admittedly was unrealistically pessimistic from the victim's perspective, but it was our belief that if the victim could retaliate despite that pessimism, deterrence should be particularly secure.

Our results indicated that strategic forces of 3,000 warheads for each side would support efficient deterrent requirements quite well, provided sufficient investment was made in protective measures. In effect this means substantial reliance on less vulnerable basing modes and on higher alert rates for bombers and submarines; both sides would have to be willing to depend on alert procedures. With that provision, deterrence appeared to be perfectly safe down to the level of 3,000 warheads, even under the more demanding theory of deterrence.

For those who believe that the threat of substantial urban-industrial damage is sufficient for deterrence, forces even smaller than 3,000 would be acceptable. At that point, however, a heated domestic debate about deterrent requirements is likely to be triggered because adopting the less demanding theory would require a change in prevailing military practice. One need not engage in that debate to undertake force reductions to the level of 3,000 warheads; reductions to that level would simply remove excess capacity.

We then asked whether these reductions, if accomplished, would remove the existing incentives for preemption and thereby establish a more robust security regime than the current one. The answer, basically, is that the reductions would bring significant improvements but that they are certainly not sufficient to establish an entirely stable international security arrangement.

There are several basic reasons why this is true, and they have to do with incentives for preemption that would not be affected or at least would not be resolved by the reductions just described. Let me briefly array those and then mention the arms control arrangements that would be designed to meet them.

First, the calculations I just summarized assume throughout that the respective command systems would operate efficiently in retaliation under the heavy damage we are assuming they would experience. This is a heroic assumption, however, and is subject to question. Unfortunately, severe disruption of a command system—though probably not absolute destruction—can be accomplished with less than 500 weapons—maybe less than 200. Even 50 weapons placed in the right places would slow the functions of the command system significantly. This means that lowering force levels would not relieve the command systems from the sense of pressure they currently face, a pressure that is probably the single most important and dangerous preemptive incentive in existing forces. The relief of that pressure would not be accomplished by the force reductions I have just described, nor would they ameliorate the problem of the troublesome engagement of short-range nuclear weapons and conventional weapons, notably in the central front and Europe. Such engagement is considered by many to be a trigger of potentially uncontrollable interaction that would not be resolved by strategic force reductions.

It is also true that the conventional balance in Europe at the moment creates pressures for preemption quite apart from the use of nuclear weapons. At least as perceived by the members of the North Atlantic Treaty Organization (NATO), Soviet forces in Europe are committed to a rapid invasion of Western Europe at the outset of war to prevent an eventual defeat imposed by superior Western economic power. The NATO perception and the Soviet doctrinal commitment produce an unstable situation that would appear to become all the more volatile as the discipline imposed by nuclear weapons is limited.

Moreover, current trends in weapons technology, if projected long enough and optimistically enough from the point of view of what the technology can accomplish, do look as if they will eventually enhance the long-range preemptive capability of even conventional forces against retaliatory nuclear forces. These trends have largely to do with the application of remote sensing, information processing, and precision guidance technology. As a result, to achieve a sufficient answer to the problem of preemption, force level reductions must be accompanied by restraint of the pace of technology or the pace of the application of technology to military hardware.

Finally, let me mention a topic that Spurgeon Keeny will explore in more detail. If partially effective defenses were deployed as strategic offensive forces were being reduced, they could raise havoc with the deterrent capacities of those forces by denying systematic target coverage in retaliation. The effect of that denial would be to reintroduce the problem of preemption, perhaps in more virulent form. For that reason, the continuation of restraints on defensive deployments and some agreed regulation of military activities in space will almost certainly be a necessary condition for strategic force reductions.

In sum, then, strategic force reductions are a promising means of protecting deterrence and controlling the pressures for preemption, but they must be combined with other measures to be fully effective: notably, with the removal of forward-based nuclear weapons in the European theater, with arrangements for stabilizing the conventional balance in that theater, with restraints on the pace and scale of force modernization, and with mutually agreed-upon restrictions on strategic defense.

Such a "full package" is what is required to produce more stable international security arrangements.

3

The Impact of Defenses on Offensive Reduction Regimes

Spurgeon M. Keeny, Jr.

The Reykjavik Summit demonstrated clearly, if there was any doubt, that strategic defenses have a critical impact on the attempt to achieve substantial reductions in offensive strategic systems. Although there were some remarkable discussions and considerable agreement at Reykjavik, when President Reagan was faced with a choice between a strategic defense and deep reductions, he chose strategic defense. In the same sense, you might say that, when faced with strategic defenses as a price for deep reductions, General Secretary Gorbachev chose not to make a deal.

It is true, as Deputy Secretary of Defense Fred Iklé and Dr. Wolfgang Panofsky have said during the seminar, that Reykjavik recorded an agreement in principle to the concept of 50 percent reductions in strategic weapons. Yet both sides actually had these positions on the table before Reykjavik. It was therefore more significant that they made notable progress in resolving some of the many difficult problems that result from the fact that the forces of the two sides are very asymmetric.

There were a number of additional issues, as Dr. Panofsky's Figure 1 indicates, that remained to be resolved. But in the final analysis, the U.S. position at Reykjavik was that in order to consent to these deep reductions, the ABM treaty should be amended in a manner that would allow unlimited development and testing of strategic defenses in the immediate future and that would convert a treaty of unlimited duration to one that terminated at a fixed date 10 years in the future. These were the points that the Soviet Union would not accept.

What was the reason for the Soviet position? Was it simply an effort to interfere with our development efforts for strategic defense, or were there rational, legitimate military concerns? Trying to analyze this problem is extremely difficult because of the confusion that exists at present as to what the SDI is all about.

As you may recall, in March 1983 when President Reagan initiated the program, he held out the vision of an essentially impregnable shield that would make nuclear weapons "impotent and obsolete." This vision remains the official objective of the program and is certainly ascribed to by President Reagan and Secretary of Defense Casper Weinberger. But beyond that, there is a vast range of views, even within the government.

For example, General Abrahamson, director of the Strategic Defense Initiative Organization, and other officials closely related to the SDI program emphasize the concept of partial defense, either as an end in itself or possibly as a stepping stone to an essentially impregnable or highly effective defense in some distant future. On the other hand, during the seminar, Secretary Iklé suggested a more modest objective: a system effective against the rather small threat that could exist when there were no ballistic missiles but only the residual problem of clandestine missiles or possibly a breakout from a treaty regime.

There are also more pragmatic approaches. The Attorney General has recently called for an early deployment with the purpose of "getting something up there," in his words, so as to lock in future administrations to a strategic defense program, whatever its objectives might be. And there is another school of thought in and out of the government that has no goal in mind but looks on the program as a bargaining chip. This school emphasizes that, to the extent we can expect any progress in arms control, it will depend on the Soviet reaction to the uncertain threat of a defense system.

I think you will find there is yet another school that looks on the SDI as part of an economic warfare against the Soviet Union with the objective of engaging the Soviet Union in a high-technology military competition. This school believes the United States has a substantial long-term advantage in this competition that will absorb Soviet capabilities in high technology and prevent the Soviet Union from becoming a greater threat to us in a broader economic sense by holding down any changes in their society that might make it a more effective competitor.

These different approaches demand very different technologies, which I am not going to discuss today. The prospects of these technologies have been addressed by others. There is an excellent paper by Harold Brown in *Foreign Affairs* (Vol. 64, No. 3, pp. 435–454, 1986) that discusses the current status and overall timetable of various technical approaches. The most recent input to this analysis is the American Physical Society report that was released on April 23, 1987. The report explores in great detail the status of the exotic technologies that were at one time to have been the focus of the SDI program.

There are obviously tremendous differences between a system that is 100 percent effective, 99 percent effective, 50 percent effective, or 10 percent effective. But it is precisely these differences and the resulting uncertainties that produce the tension between strategic defenses and real or perceived military requirements for offensive forces.

One can, for example, imagine a system that might be 50 percent effective, but one cannot assess it as being the same as a 50 percent reduction in offensive forces. The difference arises from the fact that a military planner must consider worst-case scenarios and wants to have reasonably high confidence in his war plans. If there are no strategic defenses on both sides, the military planner knows how to execute a first strike, or preemptive attack, and can make some reasonable estimates of his ability to retaliate with his secure, surviving forces after a first strike by his adversary.

But in the absence of any hard facts on the capabilities of future defense systems, a military planner will credit his opponent's future high-technology systems with potentially high capabilities. This will be true even if the military planner really suspects, or thinks he knows, that in practice the adversary's defense system would have little effectiveness and would probably collapse totally under a massive attack. A military planner concerned with ensuring his ability to retaliate after absorbing a major counterforce first strike would understandably emphasize a worst-case assessment. A nationwide defense system, whatever its effectiveness, would clearly operate best against a so-called ragged retaliatory strike, in which the retaliatory force had been reduced in its size by a counterforce attack and also reduced in its effectiveness by disorganization in timing, tactics, and possibly its target coverage.

Looking back in history, when the Pentagon planners were first reacting to the Moscow ABM system in the mid-1960s, there were estimates that if such a system were deployed by the Soviet Union on a nationwide basis, the U.S. ability to retaliate in the mid- to late 1970s might essentially shrink to zero. At the same time, I think most of the technical community assessing the Moscow system at the time had a very low opinion of the emerging Soviet defense system's capabilities, but it was impossible to quantify the system's effectiveness. Consequently, there was a tendency to assume for purposes of calculations that it just might work. This uncertain future threat had a major impact on military procurement. In particular, it was an important, perhaps the main, driving factor in the decision to go to multiple independently targetable reentry vehicles (MIRVs) to expand the firepower of our existing missile forces.

As evident from John Steinbruner's description of the CISAC study and some of our calculations, the present concept of deterrence involves a very large number of targets. Nevertheless, as he pointed out, the size of our present stockpile is so large that given mutual reductions, one could reduce 50 percent or even 75 percent of our existing strategic forces and still maintain essentially the same target coverage. Moreover, such a reduction would maintain the same extended deterrent directed not only at deterrence in the normally accepted sense but also in the sense of preventing the Soviets from sustaining their war effort in the field—on the assumption that the Soviets could do this even in the face of the loss of their society.

Some sample calculations indicated that a defense of 50 percent effectiveness would certainly not defeat an attack with the current stockpile employing the current targeting philosophy. It also could not defeat such an attack if there had been mutual 50 percent reductions in strategic offensive forces. But the problem is any system that you can credit as being 50 percent effective would be such an elaborate, nationwide system that it would probably be perceived by nervous military planners on the other side as possibly having a 90 percent capability, in the right circumstances, or even a 99 percent effectiveness, particularly against an uncertain worst-case retaliation. At these higher levels of defense, the military would not have high confidence in carrying out their retaliatory strike, given the present concepts of extended deterrent war plans.

The CISAC analysis discussed by Dr. Steinbruner essentially assumes this broader deterrent concept that, among other things, comprises attacks on a large number of military and command and control targets and on assets necessary to continue the Soviet war effort. I personally believe that a much smaller target set directed solely at economic targets would have the same deterrent significance. Such targets are essentially population targets, but as Dr. Steinbruner pointed out, the much more extended deterrent war plan also covers and has the same impact on population. In the late 1940s and early 1950s I was involved to some extent in war-plan development in the Air Force. In those days, when initially there were on the order of only 100 Nagasaki/Hiroshima-yield weapons, we thought we had not only a very powerful deterrent but that we could also very effectively impede the ability of the Soviet Union to conduct its war effort because of the high confidence that these weapons could be delivered on their targets.

With the vast expansion of the number and yields of weapons, the categories of targets that could be targeted has grown and grown and grown. Our study concluded that 100 so-called equivalent megatons would certainly be enough to inflict devastating damage on Soviet society. This is not surprising because 100 equivalent megatons means 100 one-megaton weapons or the number of weapons that would have the equivalent destructive power of 100 one-megaton weapons. Such an attack would cause some 20 million–40 million Soviet prompt fatalities, and I am sure at least twice that many delayed fatalities from untreated casualties and secondary effects. As Dr. Panofsky pointed out, a single Trident 2 submarine with D-5 missiles would have more than this amount of equivalent megatonnage on board.

This same level of damage might be achieved with the present stockpile—10,000–15,000 strategic weapons—against even a 98–99 percent effective defense. But this would involve quite a different targeting strategy in which extremely heavy firepower might be concentrated against the defended side's highest value economic/societal targets. For example, with present stockpiles, as many as 100 warheads could be directed against each major target. This would certainly accomplish the destructive objective.

Such a change in target strategy, however, would be disconcerting to the military and, again, would be subject to great uncertainties; for example, what does 99 percent really mean? Of course, the probable outcome of such a targeting doctrine would be absolutely disastrous to both countries because under that level of attack the defense systems would, in all likelihood, collapse, and there would not even be any rubble left to bounce for any of the urban areas in the two countries.

These comments illustrate the fact that a defense-oriented strategy requires not only an incredibly effective ABM system (as well as an effective air-defense system, which we have not as yet mentioned) but also radical reductions in offensive weapons as well.

As I have pointed out, however, these reductions cannot take place without the perceived loss of deterrence long before achieving a known level of defense. This is the problem of a transition point from an offensive to a defensive strategy. To my knowledge, no one has even come up with a concrete proposal as to how you walk the offensive forces down to this point while developing this highly uncertain, unpredictable defense on the other hand. This includes the President's proposal at Reykjavik, to which Secretary Iklé referred, to reduce ballistic missiles to zero.

Let me pursue this a bit further and emphasize that in approaching this transition point in which you move from deterrence to no possibility of nuclear attack, there is going to be a wide band of situations in which a preemptive strike may appear to one side, or possibly both sides, as an acceptable gamble. This would be particularly true in a situation of high tension in a severe crisis, a situation in which there appeared to be a high probability that nuclear war would occur. It might be considered an acceptable gamble if one side felt that its defense was sufficiently effective to have a reasonable chance of actually defending against the degraded offense of the other side. The force of this logic would increase if the levels of the other side's offense had been radically reduced from present levels.

I think the closest thing to any proposal that has come up has been the President's, I suppose rather offhand, idea that the United States should share its strategic defense technology with the Soviet Union. But I do not think anybody, including Secretary Weinberger, has felt that this was a credible proposal, given the present adversarial relationship of the two sides. I think the best way to describe this whole problem was put forward by the President's arms control advisor Paul Nitze, who, when pressed on this issue, said the transition problem from an offense- to a defense-oriented strategy would be "very sticky."

In this business, there is considerable experimental evidence to consider. In the late 1960s, when the United States became concerned that the Soviet Union might be contemplating a nationwide ABM defense, we did not abandon our ballistic missiles. On the contrary: we made a decision, largely driven by the potential Soviet deployment, to MIRV all of our ballistic missiles and to take other relatively simple measures, such as putting chaff on ballistic missiles to confuse Soviet radars. With those responses the military felt confident they could penetrate the Soviet ABM system.

Today, there is a fundamental contradiction in the view of our leadership on the issue of strategic defense. While pursuing the concept of strategic defense as the answer to the threat of nuclear weapons, they are extremely concerned with the possibility of what would happen if the Soviets had a strategic defense.

There are yards and yards of quotations, but I think one that gets to the heart of the matter is found in the January 1985 White House white paper on the SDI that described what would happen if the Soviet Union deployed a nationwide ABM defense: "Were they (the Soviets) to do so, as they could, deterrence would collapse and we would have no choice between surrender and suicide." A little later, Secretary Weinberger, in his famous letter to the President on the eve of the 1985 Geneva Summit, in the context of the Krasnoyarsk radar, made the statement that "even a probable Soviet territorial defense would require us to increase the number of our offensive forces and their ability to penetrate Soviet defenses to assure that our operational plans could be executed."

Now, in this whole debate the Joint Chiefs of Staff have remained remarkably quiet. One cannot help but wonder what the Joint Chiefs would say if the Soviets either had made the original SDI proposals or were to accept the U.S. proposal. If the Soviets were to say, "Let's have deep reductions and open the gates for an all-out strategic defense race," I would be most interested in how the Joint Chiefs would respond.

The uncertainty of these exchange ratios of offensive weapons in the face of an uncertain defense contrasts sharply with the rather straightforward and relatively cost-effective things that the offense can do in response to a strategic defense to maintain high confidence in its ability to deter under the broadest possible range of circumstances. I will not go into these in any detail because you are all probably familiar with them, but I will just identify the general approaches. The best way to defeat defenses is by using your existing resources to attack the extreme vulnerabilities of the defensive system. The vulnerability of prime radars was one of the reasons we abandoned the Safeguard/Sentinel approach in the late 1960s and early 1970s. Space-based defense components present the ultimate attractive targets that could cause the defensive system dependent on them to collapse, even in advance of hostilities.

Another approach is to increase firepower by building more missiles, more MIRVs, more decoys to simply overwhelm the defensive system. Another approach is just technological innovation, which can anticipate and defeat the operation of a defense system. This path leads us to such ideas as fast-burn boosters and other devices that could completely defeat, at much less cost, a very long lead-time, finely tuned defense system. Finally, there is the whole area of circumvention. If one were concerned about the ability of one's ballistic missiles to penetrate a ballistic missile defense system, one would put greater emphasis on air-launched and sea-launched cruise missiles, or possibly on other types of attack.

In conclusion, let me say that I believe that efforts to achieve a nationwide strategic defense are not compatible either in theory or in practice with a program of deep reductions in strategic offensive forces. In fact, I believe that the SDI program will, if pursued, result in an increase in the quantity and quality of strategic offensive forces.

4

The Impact of New Technologies and Noncentral Systems on Offensive Reduction Regimes

Alexander H. Flax

I would like to discuss the broader military environment in which the strategic competition exists. There is a tendency to view it in isolation; as my title implies, I am going to deal with so-called noncentral systems, meaning all nuclear and nonnuclear systems other than American and Soviet long-range strategic nuclear systems.

I will address the issues that are brought about by technological change. These issues are not all the result of radical new breakthrough technologies. More often, they stem from the kinds of technological advance that occur through incremental improvements over many, many years.

We think we have something new in the cruise missiles now in Europe. In fact, 25 years ago, we had nuclear-armed cruise missiles in Europe. The Matador, as it was called, was about two or three times the gross weight of the present cruise missile and had less than two-thirds the range. Technological improvements, mainly in the warhead and guidance accuracy, allowed cruise missiles to shrink, become more highly proliferated, and harder to verify—all characteristics that we have referred to during this seminar.

I am going to cover those aspects of many systems. I am not going to address the ABM question, but I will refer to air defense.

When we use the loose term "strategic defense," people immediately think of an ABM system and, more particularly these days, of the space-based ABM system; but in fact, there is a very large and expensive system of air defenses in the Soviet Union. These air defenses comprise thousands upon thousands of surface-to-air missile launchers and thousands of interceptors and all the radar equipment that goes with them. In a massive attack, these systems are not counted on as being very effective, but as the size of the attack shrinks, one must rethink the question.

Let me make it clear that a certain sense of unreality always obtains when one talks about large numbers of nuclear weapons. In my mind, one nuclear weapon that could be delivered with assurance would be quite a deterrent, and I note that former Secretary of Defense Robert McNamara has recently made that statement in a number of places.

That certainly was not the way things looked when we were sitting there in the Pentagon in the 1960s trying to decide what was a reasonable level of deterrence. The way in which one looks at problems outside the government is always different. It is always more detached, more rational, and so forth—and with no responsibility. One can make one's arguments and hope that the arguments will not prevail if they do not make sense.

Our discussions in this seminar involved strategic weapons reductions to the level of 6,000 warheads, which is still a very, very large number, and from there to perhaps 3,000 and perhaps 1,000 (that number has been mentioned). Let us then look at the implications. The force structure on both sides, possibly as a result of technological history, consists of a so-called triad. The triad consists of bombers, and lately bombers plus the cruise missiles; submarine-launched ballistic and cruise missiles; and land-based intercontinental ballistic missiles (ICBMs).

The triad has developed a rationale of its own: that this is a means of providing a high degree of assurance against technological and operational uncertainties and surprises, including those associated with command and control, because the command-and-control situation for each of these is, in fact, different.

Many people do not realize, however, that we have put more and more operational constraints on the command and control of bombers and ICBMs in the last 15 or 20 years, to the point that it is no longer possible, for example, for two men in a silo to attempt to launch by turning their keys simultaneously. Even with the cooperation of all the other people in their launch wing, they cannot launch because they have to have a certain code, which normally they do not have.

This safeguard, by the way, is in addition to what you have heard about the go-code. Similarly, the bombers can take off, but they cannot drop a nuclear weapon until they have that code. When the command authority decides to give them the code is another matter, but it gets to the problem of command and control, which I will say, without attempting to justify it here, is most severe for the most survivable weapon, the submarine-launched ballistic missile. When balancing up the forces, when making reductions, one has to think about all of these aspects.

In addition to the difference in the way these triad weapons survive, there are also differences in the way they reach their targets. The bombers and cruise missiles must go through the opponent's air defenses, and the ICBMs and sub-launched ballistic missiles must go through ballistic missile defenses, if there are any.

As it happens, there are some ballistic missile defenses in the Soviet Union, but they do not exceed 100 interceptors and thus are permitted under the ABM treaty. Again, at the levels we are talking about here—that is, thousands of warheads—these defenses hardly make any difference. If you go to shrinking down to smaller and smaller numbers, the so-called ragged attack, it would matter—or could matter—because certain selected targets could be protected.

There are certain possibilities that people argue about: unanticipated technical advances in submarine detection; anticipated improvements in guidance accuracy, with the result that hard silos could be destroyed even with nonnuclear warheads; failure of the ballistic missile warning system on which the bombers depend. The point is, we have assumed that these will never all happen simultaneously; thus, the insurance factor is there. As you reduce your forces, however, that factor has to be looked at very carefully.

The main point I want to get to is that the numbers being discussed are total numbers of warheads. Because of operational and economic factors—I will make a rough estimate, and there are refinements one can make—really, only about half of those can be counted on in a retaliatory strike because of the different alert rates of bombers, submarine-launched missiles, and land-based missiles.

Thus, when we talk about 3,000 warheads, we are actually down to 1,500 that would be available for retaliation. As one gets down to those levels, we may be talking about having the entire ballistic missile submarine force consist of about six to eight submarines, of which three to six might actually be out on patrol at any given time.

In this case, we would be putting our reliance on the deterrent in those small numbers—that is, if we keep to the present concept of Trident subs carrying 24 missiles and 7 warheads per missile, with the terrible destructive power that has been described. Actually, this is the most economical way to do it. You may have noticed this during the debate on whether Midgetman, the proposed small, mobile ICBM, should have one or three warheads. The whole controversy was largely an economic argument.

The economic way, then, is to go for economies of scale. Yet from the standpoint of assurance, robustness, survival, you do not want to do that. For example, similarly, at this level of 2,000–3,000 warheads, we might have 15 bombers on alert, all told, out of a total of 45 bombers.

The problem that I am going to focus on in the remainder of this discussion is that the noncentral systems, particularly with technological improvements, improvements in structural weight, propellants, guidance accuracy, and so forth, have residual strategic possibilities. Many of these are dual-purpose systems, even ballistic missiles—you may note that ballistic missiles fly between Iraq and Iran these days with conventional warheads.

For example, consider the Soviet SS-20 intermediate-range ballistic missiles (IRBMs) with three warheads, that we are talking about phasing out in an INF agreement. If we take two of those warheads out and leave one, the missile almost has intercontinental range—that is, it is now a single-warhead ICBM. Thus, it is very important to eliminate those if you are talking about deep cuts.

Similarly, all military aircraft—including all of our fighters—are tanker-refueled many times when they are flown to Europe. That is how they get there. Thus, they have intercontinental range. You may have noticed in the raid on Libya out of the United Kingdom, for example, the F-111s were repeatedly refueled to go all the way around the Bay of Biscay and into the Mediterranean because they could not cross France.

You may also have noticed that the little Voyager that flew around the world without refueling was only a 12,000-pound vehicle. Now, admittedly, it flew at very low speed, but its achievement shows the potential in some of the more advanced structural composite technologies and also in our propulsion technologies. Thus, we have to worry about intermediate-range missiles, if they exist, when we get down to small numbers.

We also have to worry about tactical aircraft, and even tactical transport aircraft like the C-141, C-5, and the C-130, which has a shorter range. All have cargo doors in the rear that open for parachute extraction so that loads can be dropped to troops in the field. Those doors are also very good for extracting cruise missiles. They are also good for extracting anything else. Potentially, they can become bombers without much physical change.

My argument here is not that this all makes a treaty impossible but rather that a much more intrusive, more widely encompassing verification regime is needed. I might also say that even space-launch vehicles, which originally were all converted ballistic missiles, are perfectly good ICBMs. And one can test all of the elements except the reentry vehicle by conducting a space launch.

Space launchers thus are another worry, and here there is an asymmetry between ourselves and the Soviet Union. Some people say that the Soviet space program is better and bigger than ours because they have 100 launches per year to our 10 or 12. Yet their spacecraft do not last very long in orbit. Nevertheless, they have large production lines and big launch facilities for space launchers, which, although not very good for retaliation, are quite effective if the first strike is the name of the game. Again, intrusive inspection is going to be very important in this area.

Modern cruise missiles such as our Tomahawk, for example, have both conventional and nuclear roles. The antiship missile—the same missile, practically—carries a much bigger conventional warhead to put a hole in a ship with armor. As a nuclear delivery vehicle, it can suddenly expand its range by a factor of about three and fly from Europe to the Soviet Union.

Years ago, we did have cruise missiles that flew intercontinentally. They were called Snarks. They were not very good, but with modern technology, they too would be possible candidates for a nuclear role. Again, they are small, they are difficult to find, and they can be tested on a racetrack instead of over full range. There are all kinds of problems that argue for a much more cooperative and highly intrusive relationship in verification.

The other thing I want to address from the standpoint of technology is the survivability of the land-based force. I am not going to go into all the potentials for antisubmarine warfare (they are highly complicated and mostly classified). We and the Soviets, instead of relying on ICBM silo hardness, are increasingly "going mobile," either by road or rail, and both countries either have in process or are planning ICBM systems of that kind.

It is easy to say we will put mobile ICBMs on this much territory, and they will wander around, they will not be seen. Certainly, in the present situation, our overhead systems and our command-and-control systems could not do a credible job of real-time retargeting. It is almost certain, however, that in 10 or 15 years, they will be able to do that. Thus, we will have that problem, that future window of vulnerability for the mobile land-based system, and what we will do about it is not at all clear.

All of these problems must be considered. We cannot write treaties with vague, ill-considered clauses as we did in the first strategic arms limitation treaty (SALT I) (and even in SALT II, in some cases) and then worry about the problems later. Such a process creates ill will, fear, and suspicion; it really is counterproductive. It is really quite important to recognize some of these factors as we move in the direction of deep cuts in strategic offensive forces.

Because of the relatively small numbers of launch platforms that may be involved as we go to 3,000 warheads and below, we really have to consider modifying our launch platform concepts. We probably do not want Trident submarines carrying 24 missiles. We probably do not want big bombers carrying 20 cruise missiles. It is not going to be the most economic approach. It could be that we will end up with a force of one-third the size of our current force, costing roughly what the present strategic force costs. Nevertheless, everybody should realize that that is a problem that needs to be considered.

One final point: I want to comment that I am not unmindful of the problems of third-country forces. I deliberately excluded them as a limiting factor because I think some of these other things will come into play even before the third-country forces.

5

Alliance Issues

Catherine M. Kelleher

I am pleased to discuss the issue of deep strategic arms cuts as it affects not only NATO and its future but also European attitudes on defense and arms control more generally. Of all the topics touched on in the seminar, this issue and the comments of Dr. Paul Doty on the conventional balance go to the heart of the political dilemmas in the arms control regime discussed at Reykjavik. In the long run, these, indeed, are the issues that will have the greatest effect in terms of restructuring the international order.

In this sense, we find ourselves precisely at the point Dr. Panofsky outlined earlier. We know that something new is happening and that something different will result. But we have not yet thought through all of the implications or, indeed, all of the consequences for many of the safe assumptions and easy premises on which we have based much of our defense effort below the strategic level in the postwar period.

Specifically, in terms of Western Europe, there is no other area in the world beyond the homelands of the United States and the Soviet Union for which a deep-cuts regime will have greater political and military implications. This is true not only because of extended deterrence and the U.S. guarantee but also because in Western Europe there is a military force, significant and nuclear, that is unmatched anywhere else in the world. It is composed not only of a substantial number of American and Soviet forces, but also a rapidly increasing number of independent French and British forces.

Yet what is perhaps the most difficult part of this short discussion is not so much the technical assessment of these forces or, in the tradition of CISAC, a better technological base for searching for an arms control regime. What is being sought is a political solution to basic questions of political trust and the degree to which perceptions shape reality, however much technical calculation may end up with a somewhat different outcome.

A recent Herblock cartoon (p.36) questioning what the Europeans really want may be a good starting point. In part, I think it reflects the prevailing perception of publics and elites here in the United States that the Europeans are never satisfied; that no matter what one is talking about, there are at least contrary, if not critical, European views of what the United States is attempting to achieve.

It seems to me this comes from a set of fundamental paradoxes that are not new that have been present in European thinking about defense and deterrence, indeed, since the first hours after Hiroshima. These paradoxes will come to be of central importance in our discussions and deliberations of a deep-cuts world much sooner and with more intensity than even the present discussions about INF suggest.

In this discussion, I will focus on only three of these paradoxes. The first follows, perhaps, from what I have just said: namely that whatever the issue they are confronting, the Europeans are basically the primary status quo powers in the world today. Any change in the political and military framework in which they find themselves is itself threatening. This is true even if the long-term outcome may actually be even less risky to them—in terms of, let us say, the deployment of nuclear weapons on the ground in Europe or a regime of deep arms cuts that makes the United States less vulnerable to Soviet ability to preempt or to launch a disabling strike.

A second paradox is that the NATO regime at the moment is based on a system of extended deterrence and on an American

From *Herblock at Large* (Pantheon Books, 1987).

nuclear guarantee. The assumption, at least doctrinally, is that these weapons will be used according to the principles of the flexible response strategy adopted by NATO in 1967.

Yet if we look at particular force postures or at expectations about the connectivity of political and military leadership during a crisis, it becomes very clear that extended deterrence—and the NATO system so far—are based on two quite contradictory assumptions.

The first assumption is that there is little, if any, threat of war in Europe at the moment, and therefore a deterrent system is adequate. The second assumption is that in case of a threat of war, the very inadequacy of forces, and of political control at all levels below central strategic systems, will make the threat of initiating escalation to an intercontinental exchange, and thus to homeland damage, quite credible.

Let me put forward yet another dilemma. Much as George Bernard Shaw said about Christianity, the alliance is basically unsatisfactory and proves with every passing year increasingly unsatisfactory in terms of a natural convergence of European and American attitudes and agendas. Particularly in the last 10 years, Europeans have become increasingly frustrated with what seems to be an on-the-job training program for American presidents. They also see Washington as acting on a questionable set of assumptions, whether it be in the raid on Libya or at Reykjavik itself, and overlooking what the limits or the possibilities for alliance consultative practices really are.

In the view of many Europeans—and I will be more specific later about how Europeans differ here—the United States is capable of doing anything that occurs to it upon arising one morning. Yet in the views of all of those who are engaged even in the much-intensified staff talks and national talks among European governments, all other alternatives to the present form of the Atlantic alliance are either not available at the moment or are in fact vulnerable to the same types of contradictory political, economic, and even military pressures that led to the political instability characteristic of Europe in the 1920s and 1930s.

These dilemmas are known and describable, yet they do not seem susceptible now—nor, I suspect, will they be susceptible for a long time—to satisfactory resolution, either in terms of logic or in terms of predictable political outcomes. Let me turn therefore to the

European reactions to the specific deep-cuts regime that has been outlined in the discussions by Dr. Steinbruner and Mr. Keeny.

One must ask here if there are, in fact, inherent difficulties within that regime at the 50 percent level of cuts that would suggest a point at which Europeans would wish to do something else. This "something else" might involve the development of weapons systems of their own or perhaps a movement toward the formation of new political frameworks in which to seek their own security.

I think that, within the limits of the foreseeable, there is strong support for superpower arms control on the part of most Europeans at both the conceptual and operational levels. Arms control would be looked on as a step toward superpower rationality and a step away from what seems to be the greatest threat for those on the left in Europe: namely, a conflict based in Europe in which the superpowers would drag the Europeans into a fight between them.

There would be no particular worry on the part of Europeans at the level of 50 percent cuts, or even, I suspect, at levels of 70–75 percent cuts, about the adequacy of what would remain for deterrent purposes. And this confidence would probably hold true no matter what existed in terms of deep concerns about the U.S. promise of extended deterrence and its foreseeable operational significance.

In fact, in the view of most Europeans, deterrence is a question of uncertainty. It relies on the premise that neither side can be sure that the other side would not do something stupid or would not do something that was not calculated (or calculable) in advance. The 30 percent of strategic forces remaining after the cuts is seen as more than adequate to instill such a fear.

What Europeans worry about are two other factors, and here we ascend (or descend) into the political realm in which the technical assessment becomes less important. First and most important (one saw it in the weeks right after Reykjavik), Europeans are concerned about multilateralizing the arms control process; at the moment the superpowers seem to be deciding the central issues of security for them without much consultation. To some extent, they are afraid, as always, of a superpower condominium concluded at their expense and over their heads. Given their fears about the general and perhaps growing instability within the American political system, they also believe that such a condominium would merely be the first step on a slippery slope that might take them to many other places they did not wish to go.

Quite specifically, this slippery slope may well be one that leads to a superpower agreement on the denuclearization of Europe. Yet such an agreement might involve significant threats and risks of conflict and would make Europe safe for either conventional or limited nuclear war while the superpowers remained sanctuaries safe from the battle and able to pick up the pieces at the end.

Next, what is most difficult to remember in terms of the present INF discussions is that there will still be significant nuclear forces left in Europe (Table 1). Thus, it is a question of stabilizing relevant weapons balances and preserving the rights of deployment, modernization, and control, while at the same time avoiding any increase in the probability of confrontation or conflict in Europe itself.

Most of us tend to focus on the models of the strategic exchange and forget the precise numbers of forces and the kinds of forces that exist within Europe. It is particularly interesting when one considers the numbers of medium-range missiles, as Dr. Flax has reminded us, that can, in fact, receive both an extension in range and a change in destructive capability simply by changing the warhead or a propulsion system.

The top category on Table 1, the medium-range missiles or INF, is what is currently under negotiation. But I would also direct your attention to those that remain even after a U.S./Soviet agreement—that is, the British and French forces, which are not many but still enough to constitute a sizable force. There are 226 warheads on the present French and British systems, a number that will increase threefold when the final stages of modernization, the so-called third-stage development, of the British and French forces is completed in the mid-1990s.

Shorter-range systems are now being talked about in terms of the second zero-zero agreement, the add-on shorter-range intermediate-range nuclear forces (SRINF) discussions. There are very few on the side of the United States at the moment: only the Pershing 1 launchers, which are West German in ownership but for which the United States maintains the warheads in stockpiles on West German territory.

At the moment, as far as we can tell—and here we enter the world

TABLE 1 Current Deployment of Nuclear Arms in Europe

Weapons	United States	USSR	U.S. Allies' Weapons
Medium-range missiles (1,000–3,000 miles)	316 Warheads: 208 cruise missiles, 108 Pershing 2s	922 Warheads: 810 on SS-20s and 112 on SS-4s[a]	226 Warheads: 80 on French M-20s; 18 on French S-3s, 64 on French M-4s, 64 on British Polaris
Shorter-range missiles (300–600 miles)	None	About 80 missile launchers in Europe	72 Missile launchers in West Germany[b,c]
Battlefield-range missiles (less than 300 miles)	108 Missile launchers	About 1,100 missile launchers	99 Missile launchers
Aircraft capable of carrying nuclear weapons	About 500	About 2,000	About 1,300
Nuclear artillery	About 1,000 launchers	About 3,600 launchers	About 2,000 launchers
Warheads (est.)	3,500	3,000–7,000[d]	1,700

[a]This is a NATO estimate; the Russians say they have 729 warheads.

[b]Nuclear warheads controlled by the United States.

[c]The West German government has stated that the 72 missile launchers in West Germany will be phased out once the U.S. and Soviet medium- and shorter-range systems are dismantled under the terms of the INF agreement currently nearing completion.

[d]This is a rough estimate.

Adapted from: *The New York Times,* April 23, 1987, p. A6, © 1987 by the New York Times Company. Reprinted by permission.

SOURCE: International Institute for Strategic Studies; United States government (numbers for Soviet shorter-range missiles and battlefield-range missiles).

of intelligence about Soviet tactical nuclear capabilities in Europe—there are probably something like 80 missile launchers of the SS-23 or SS-12/22 type existing in Western Europe at this time. Without an agreement, most analysts expect a substantial increase in these numbers.

Proceeding down Table 1, we have battlefield-range systems—those with ranges of less than 300 miles. Here we are primarily talking, on the American side, of the missile system known as Lance, and eventually the system known as J-TACM, which will be deployed. On the Soviet side, we are talking about the Scud-B and the SS-21. Here, too, there are missile launchers primarily controlled by Britain but also some French missile launchers that will come into play.

The next category in which the numbers, I think, are really quite striking is that of aircraft capable of carrying nuclear weapons (i.e., capable but not necessarily certified). On both sides, there are in the range of 1,800–2,000 aircraft that could carry a minimum of two bombs at any one time.

With nuclear artillery, we are talking about 3,000–3,600 launchers. These are mostly of the 8-in., 155-mm variety, and would be capable of being equipped with a nuclear warhead as well as with conventional munitions. In terms of the number of warheads that are involved, we know fairly well what is to be counted on the Western side. There are currently 3,500 warheads held by the United States for its own use in Europe and 1,700 other warheads that are either maintained for Allied use or maintained independently by the French, and to some extent by the British.

The Soviet figure is extremely soft and really represents a best "guesstimate." There is a considerable range of estimates about what the Soviets have, not in Eastern Europe itself, but in the Western military districts of the Soviet Union, which are probably reserved for movement forward in times of conflict.

When one goes beyond looking at these particular capabilities and considers the guarantees Europeans will want to have for the coming balances in these forces, one encounters significant distinctions among the different European positions.

There are three important dimensions that should be taken into account. The first is the very wide gap that exists between public attitudes and elite opinion in every country but France. Here, to a large extent, one is paying the cost of several decades of irresponsibility on the part of European political elites who have never found the "moment opportune" or who have never been faced with the necessity of explaining precisely to their publics what it is that they have been agreeing to within NATO, much less what they consider to be necessary in terms of national security.

For most European publics, war in Europe is unthinkable and unlikely. And for elites, it is unthinkable, but only so long as deterrence based on the theory of nuclear risk and the American guarantee, however thin and threadbare that now is, is maintained.

For European publics, nuclear weapons are the problem. Therefore, denuclearization as soon as possible appears to be a solution that would appeal not only in the short term but indeed, in the very long term as well. For elites, on the other hand, there is a consistent tendency across countries (perhaps with the current exception of the Scandinavians) to look at a total mix of forces. This mix would require some nuclear weapons on the ground in Europe as necessary elements in the total balance of conventional and nuclear forces.

Small numbers do count, then, if only because of the softness and the high degree of congestion in Europe and the vulnerabilities of major urban and economic capability targets. For the elites, as opposed to the publics, any step to be taken toward limiting these weapons would have to involve an adequate, credible verification scheme and a comprehensive but not necessarily symmetrical regime of East-West reductions.

A second major dimension on which there is debate involves the split between the nuclear European nations and the nonnuclear European nations. For the nonnuclear European nations (here, I think, one hears the most in this country about West Germany but it is also true for Belgium, the Netherlands, Luxembourg, and the Scandinavian countries), nuclear weapons are, in fact, of questionable legitimacy. In their view, such weapons may be necessary, but there is always the question as to just how they will be used and with what legitimacy. The major item on the agenda of these countries is to control the actions of nuclear states, both allies as well as adversaries, through arms control or even the mechanics of the North Atlantic alliance.

Here, however, another paradox arises. Conventional defense

efforts, in the views of these countries, are at present at the maximum they can be or will be for at least the coming two decades; that is, the maximum that is possible, given European demographics, and that is tolerable in political and economic terms.

For the nuclear states, nuclear weapons are necessary and legitimate. They are the first priority in terms of national autonomy, an autonomy that will be even more highly prized in a world of deep cuts. Conventional defense efforts are important but receive nowhere near the same priority, either in terms of public attitudes or the attitudes of competing conceivable elites.

The last dimension is the real difference among the positions of the Federal Republic of Germany, in terms of its requirements for both defense and deterrence, and those held by other Europeans. What is involved here is the preference of the West Germans for an American guarantee under almost all imaginable circumstances. In part, they have traded the acceptance of a nonnuclear status for that guarantee and for the promise that there would be the forward defense of all West German territory; that is, the defense as far forward as the demarcation line between East and West Germany as specified in the agreements that brought West Germany into NATO in the mid-1950s.

This trade is still one for which West Germany expects to pay in terms of loyalty of a kind and alliance membership and for which it expects the United States to continue to make the same kinds of guarantees, even in a world of uncertain deterrence. It is a political task, but it would also be the task of any alternative defense arrangement other than the NATO alliance that can be imagined. The forward defense of West German territory, which involves forces other than simply West German forces, thus becomes a sine qua non for West German participation in any scheme.

For the other Europeans the idea of a transition to a supplementary European defense arrangement, particularly one in a deep-cuts world, is, in fact, more and more an idea whose time may yet come. But most probably it would be a European defense arrangement based on Franco-British cooperation, the third-generation nuclear forces of both being tied together and subject to the control of the other Europeans. There would also have to be a different kind of political cooperation and coordination with the United States.

Involved in this, however, is what precise role West Germany would play and, in particular, West German demands for this kind of all-encompassing forward defense. Despite the fact that 40 years have elapsed since the end of World War II, the idea of a nuclear-armed West Germany is still anathema to most Europeans. It is the last step to be contemplated and only then after U.S. withdrawal and the development of European defense arrangements and other "fixes" for West Germany's security needs.

What is the conclusion of this discussion? It seems to me that what we are looking at is a number of things that "do not compute." One adds up the columns and comes to the conclusion that some compromise, some "give," some political solution, will have to be sought, and for this there is the all-important and all-purpose requirement of leadership, a leadership not yet visible, at least in the present discussions on INF.

Yet there is a conclusion and set of points for short-term policy guidance that, it seems to me, do emerge out of all these political calculations. The first and unquestionably the most important is that if U.S. strategic forces are seen as decoupled from European defense, the result is unacceptable. No scheme of superpower cuts that makes these tactical nuclear forces the central elements of the continuing system of deterrence, thereby making Europe safe for limited nuclear war or for a protracted conventional war, will be acceptable to European elites or publics. This situation is also not conducive to the continued health or existence of NATO as we now know it.

Second, in the formulation of a deep-cuts regime, one must certainly be guided by one clear lesson from Reykjavik: the process by which deep cuts and any other associated constraints are undertaken must be quite careful and very consultative, unlike Reykjavik. At the moment, it appears that the United States does consult in the sense of giving information and listening carefully, although not conclusively, to the demands of its allies.

For the desired political and military effects within the new order that is suggested by a deep-cuts regime, one must posit, at least, the continued existence of a stable, peaceful political order in both Europes.

Last, but not least, the question of independence and autonomy is not one that is going to be solved simply by allowing the size of French and British forces to be noticed around the third stage in a deep-cuts regime. At some point, it has to become clear that these independent nuclear forces in fact assume a greater and greater role, not just for themselves but in terms of the expectations others have about their willingness to abide by the general rules of the game in the political order that will follow.

Finally, perhaps the best longer-term prediction is that no matter how deep the cuts, no matter how deep the set of deliberations we will have now, there will still be plenty of work to be done in the discussion of European attitudes on defense and arms control.

6

Implications for Conventional Forces

Paul M. Doty

In my many excursions from the roles of chemist or biochemist, I have dealt mostly with nuclear problems, as has CISAC itself; but with the advent of the possible elimination of long-range and shorter-range nuclear weapons in Europe, one must ask how this affects the conventional force situation in Europe. That question, in turn, causes one to ask what that situation is and how much it can be and should be changed in response to these changes in nuclear deployments.

I wish the answer were simple so that I could be brief, but the situation is extremely complex, as Dr. Kelleher has pointed out. It is a much more complex situation than the strategic one because there are many more factors involved.

The categories of conventional weapons, such as main battle tanks, do not have the same consistent meaning as do missiles of a given size and range. The interdependence of the factors—the roles of leadership, readiness, location, and logistics; the maintenance of the equipment; the supply of ammunition; the motivation of the troops; the involvement with civilians; the state of communications, command, and control—all of these form a complex, interacting scene that is unfathomable if one is looking for a simple answer.

The problem that we really face, then, is to understand the inexactitude of the conventional situation and ask whether that uncertainty is so great that the proposed changes on the nuclear deployments are small in comparison, or whether the opposite applies.

Indeed, to speak of the balance of forces propels us into the sin that Winston Churchill called "terminological inexactitudes" because a balance of forces implies a rather simple weighing of objects on two pans of a balance. What we have instead is literally dozens of factors of different weight, the weight of each one depending on how it interacts with the other. Difficult though it may be, such complexity must be addressed.

At this point, it may be wise to supplement somewhat Dr. Kelleher's discussion of the situation in Europe. We formed NATO in 1949; the Warsaw Pact organization was formed a year later. In the 1950s the decision was made to rearm West Germany but only with conventional arms—West Germany was not to have nuclear weapons. In the early 1960s, under the influence of Charles de Gaulle, France withdrew from military coordination with NATO, an action that continues to be a loss to NATO's effectiveness. Indeed, in those early years many issues were decided that are still with us today.

Perhaps the most important one was the fact of that time that nuclear weaponry was an easy and cheap substitute for manpower. Thus, with the introduction of tactical battlefield weapons in Europe in the late 1950s, there was a ready acceptance in all quarters of the proposition that one did not need the massive armies in the millions that had characterized World War II. That proposition is now questioned in many ways, but it is still a historical fact that casts its shadow over all that we say.

What are the goals of NATO? Very simply, its goals are to discourage a Warsaw Pact attack and to buy time after one occurs with which to wrestle with the nuclear decision. We are happily aware that for more than 40 years war in Europe has been deterred. Most people think the existence of nuclear weapons has been the major deterrent to war, and that probably does rank first; but there are two other possibilities.

One is that the conventional forces deterred attack, and the other is the related possibility that the Soviets did not find it in their interest to attack. The theology of deterrence has gone through many loops, and none is more clear than the one that faces us today with the possibility of the withdrawal of the 1,700 or so missiles in Europe of intermediate range.

The arguments that went into the decision to deploy missiles on the NATO side in 1978 and 1979 were of two principal types. One stated that the missiles were the counter to the large number of SS-20 missiles that the Soviets had been deploying during the 1970s; the other was that the missiles were needed for coupling the American strategic forces to the defense of the continent.

This dilemma is with us today, and it is being argued in *The Washington Post* between Paul Nitze and Brent Scowcroft as well as in op-ed pieces and in many other arenas.

The Soviet offer to accept the original U.S. proposal of 1981 to have zero missiles of this size on both sides has reinvigorated the arms control negotiations. To follow that offer with the proposal that they withdraw their 130 weapons of shorter range—whereas we have none—has compounded our surprise: The Soviet proposal was quite unexpected, and there has been a great nervousness in NATO circles as to how to respond to this seeming largess. The question seems to be this: Would "double zero," which is the code word for going to zero in both shorter- and intermediate-range missiles, mean changes in conventional forces, or do we need such changes anyway and does arms control have an important role to play here?

Finally, I think we must always bear in mind in these considerations that nuclear weapons, of whatever size and number, are not the only deterrent against Soviet attack in Europe. What is more important, particularly in coupling the defense of Europe to the American strategic arsenal, is the presence of 330,000 American troops there. This is so well known, such a commonplace, that it is almost forgotten, but it is these U.S. troops that are the heart of the deterrent. One cannot imagine that a country would give up that many of its youth without a retaliatory act of some kind.

Let us consider, then, the conventional force situation in Europe. One might call this exercise "bean counts and scenarios" because there are two general approaches to evaluating the effectiveness of conventional military forces. One is to count whatever there is to count and see how it adds up. The other is to say that, even if you do that well, it is not enough, by any means, because what really counts is how these forces would behave under typical, imaginable conditions of confrontation and war. In this latter case, one goes on to identify the most likely scenarios of conflict and how the forces on both sides would interact under those conditions.

First, however, let us go back to the bean counting. The unit of conventional forces is divisions. Divisions are themselves quite inexact, but on the Western side they number about 15,000 troops plus some civilian support. On both sides they are in varying states of readiness, which are designated as category 1, category 2, or category 3.

Categories 2 and 3 are not filled out with respect to manpower. Most of their weapons are stored. These troops do not exercise very much, they often do other jobs, and they cannot be in any sense counted as category 1, which means forces that are pretty well prepared to do battle. ("Pretty well prepared" can be defined as ready to move to a designated location following a very short period of mobilization and practice, maybe only a few days.)

If we look only at the category 1 divisions, then, on the Warsaw Pact side there are roughly 50, and on the NATO side there are roughly 36, including 10 French divisions, 7 of which are in territorial France. Thus, our numbers are 50 against 36. That seems a little uneven, but it becomes more balanced when one recognizes that divisions in the Warsaw Pact are considerably smaller than they are in the NATO countries and one is not far off to assume that the NATO divisions are 50 percent larger.

If one corrects for that, then the 36 becomes 52, and the numbers of equivalent category 1 divisions across the 500-mile inter-German border are about equal in size. Many people would argue that the quality, reliability, and state of readiness of the Allied divisions is greater and that the Warsaw Pact forces are ahead in numbers of weapons and ease of resupply. Other factors come in as well, but I think that that is a general judgment.

The category 2 and category 3 divisions, which are, for the most part, farther back (even as far as the continental United States and beyond the Urals in the USSR), are much more numerous on the Warsaw Pact side. If they were brought up to the strength, readiness, and equipment that category 1 divisions have, they would make the unevenness much greater.

The ultimate deployment of all three of these types of divisions total about 110 in the Warsaw Pact and 49 for NATO. A consequence of this is that the balance is not bad for the first days or even weeks of a war; in the first few months after that, however, the more numerous category 2 and 3 divisions of the Warsaw Pact could be brought up to strength. The odds, then, are very much against the West in a conventional war.

Many other factors enter into the evaluation of these numbers. I mention only one (perhaps the most important and the most neglected), and that is the long-held, traditional military judgment that defense requires less manpower than offense. The existence of prepared positions for defense, the knowledge of the territory, all of that introduces factors that are never known but are often judged collectively to favor the defense by as much as three to one, or at least two to one. In military terms this is a substantial factor that weighs quite favorably on the NATO side.

If we turn now from manpower to equipment, the situation is more favorable in the bean count to the Warsaw Pact side. For main battle tanks, the most common element of military equipment, there are about 2.3 times as many on the Warsaw Pact side as in NATO; artillery, 2.7 times as many; armed helicopters, 3 times as many; and antiaircraft guns, about even. Thus, if one goes through the equipment inventory, there is no doubt that, in terms of numbers, there is more on their side than on ours in most categories. However, the Warsaw Pact equipment is, on the whole, older, less well maintained, and less transportable, and gives rise to a lower rate of fire. The imbalance in military equipment, then, is a factor, but it may not be a decisive one.

William Kaufmann, now at Harvard, has estimated Warsaw Pact effectiveness, under various assumptions, in an engagement on the central front. Under a conservative set of assumptions, including current capabilities, he concluded that the probability of a breakthrough after a Pact attack is about 50 percent in the first 10 days and rises to 75 percent in subsequent weeks and months. This case, however, includes the assumption that the non-Soviet Pact forces, which make up about one-third of the Warsaw Pact forces, operate at peak effectiveness. If the opposite assumption is made, that they contribute only negligibly, then the probability of a breakthrough diminishes to about 22 percent in the first 10 days but rises to 53 percent in 2 weeks and to 65 percent in 3 months. Again, if there is time for the entire Soviet military to concentrate and move against Western Europe, the chance of a breakthrough increases above 50 percent. Although such estimates are sensitive to many assumptions, they do convey the central point: the present balance is such that the Warsaw Pact could not realistically plan on success—at least in a few weeks—if it initiated war.

Finally, if we consider what NATO needs to be more effective militarily, there is a well-examined list of items, none of which involve adding more divisions. For example, "tactical air" is the term used to define the close air support to ground troops. Many people argue that our air superiority is canceled by the dense air defenses in the Warsaw Pact countries and the large number of Soviet interceptors. Others would argue, however, that the superiority of our individual planes and the better readiness of our pilots would more than compensate for this and that, indeed, one can expect, in an engagement, air superiority to belong to the West.

The problem is that perhaps too many of the NATO aircraft are devoted to long-distance interdiction and fighting over the skies of Eastern Europe and not enough to helping the troops on the ground. One rather common proposal to improve NATO effectiveness is to provide more air support—planes that are coupled to the combat on the ground.

An equally large item is "smart munitions," which are entering into their second generation now and are increasingly spectacular in their demonstrations. The large numbers of tanks that we have so feared are becoming increasingly vulnerable to antitank weapons of growing sophistication. Our command and control has also improved greatly, but it would always be highly vulnerable if the war turned nuclear.

Other high-tech systems are on the verge of being deployed—

that is, if we wish to pay the bill. Perhaps the most spectacular is called J-STARS, in which planes of the 707 type, carrying enormous amounts of radar, fly north and south some miles inside West Germany, from which one can "see" 200 miles into East Germany and even the edges of Poland. From there, the system can detect, for example, the difference between a tank and a truck, follow their movements, predict their speed, choose what system should attack them, give the orders, and guide them there.

There are more items on this NATO improvement list, but I will mention only one more, an item whose time may come although it has been intensely resisted by the West Germans. This plan is simply to erect an effective barrier (which could be forests, ravines, or other places for the ready emplacement of mines) against tanks and other conventional forces along the western side of the inter-German border. One could, at relatively small expense, create an extremely efficient barrier there, greatly improving the effectiveness of all of the other parts of the forces at very low cost and with practically no added manpower.

I present these options in this much detail merely to pose the notion that if one wishes to invest further in NATO forces, it is not obvious that adding divisions (in order to correct the numerical balance among the forces) is the right thing to do. There are many quite intelligent changes to be made that are not dependent on adding divisions.

What, then, are the conclusions on the conventional military balance? These are mostly a matter of judgment, and other people will probably give other answers than mine. Nevertheless, I would list them as the fact that the unfavorable ratios on equipment may not be decisive and that our ability to go the high-tech route earlier, faster, and more effectively than the Warsaw Pact decreases the need for additional weaponry. On the average, a smart weapon is 20 times as efficient as a "dumb" one and, therefore, the counting in the future has to take that into account. More divisions for NATO, then, are not necessarily its greatest need.

There is another general conclusion that I hope is implicit in what I have presented; that is, the uncertainties in estimating NATO capabilities are far greater than the effects, for example, of changing the total of Soviet nuclear weapons that could be targeted on Western

Europe by less than 10 percent, which is what would be done under an INF treaty. I think the same follows for the 50 percent cuts in strategic forces. Thus we have a situation here that, as Dr. Kelleher pointed out, is not at all satisfactory from many points of view but is almost invariant to changes of the kind that are being discussed in nuclear arms control.

The Institute for Strategic Studies in London has the longest history of consistent examination of this problem, and its most recent conclusion is that the balance is still such as to make general military aggression a highly risky undertaking for either side. The initial advantage to an attacker is not sufficient to guarantee victory; also the consequences for an attacker would still be quite unpredictable, and the risks, particularly of nuclear escalation, remain incalculable.

If this could be put in quasi-percentage numbers, we might say that the Europeans would like to have at least a 90 percent certainty that the United States would use nuclear weapons in their defense. They feel that the likelihood is much less than that, maybe actually around 40 or 50 percent. Yet I believe the Soviets would not dare risk testing this situation, even if the likelihood were as low as 5 percent.

Against this admittedly selective examination of conventional forces, what is the outlook and possible role of arms control? There have been under way, unknown to most people, I think, two substantial attempts at arms control in the conventional force area. The first might be called a structural approach because it deals with the structure of the forces (manpower, units, and equipment).

The prime example of this approach has been the Mutual and Balanced Force Reductions (MBFR) negotiations in Vienna that have been going on since 1973 without a product. The aim in this case has been to use manpower as the important variable and to find ways of reducing that. It has failed largely because it has not been possible for the two sides, the two blocs, to agree on the existing manpower count, and the difference is a large number—more than 100,000 troops.

Part of this problem is due to the fact that troops are defined differently on the two sides. For example, there are construction crews associated with the Soviet armed forces that mostly do construction and work of that kind, but it has proved to be an insoluble problem with respect to a common definition.

One may think of these 15 years as being wasted, but in point of fact, many subsidiary and important problems have been worked out during this period. Nevertheless, the negotiations have no product, and interest in this area was revived only by a speech of General Secretary Gorbachev's a year ago, in which he called for substantial reductions in all components of land forces and tactical air forces from the Atlantic to the Urals.

Such a proposal certainly woke everyone up because if he meant what he said, substantial reductions, much greater than the numbers being bandied about in Vienna, are negotiable. In addition, enlarging the area affected from the Atlantic to the Urals brought in the whole western Soviet Union, thereby reducing the logistical advantage of the Warsaw Pact.

General Secretary Gorbachev added that tactical nuclear weapons should be removed at the same rate as conventional weapons and that onsite inspection should be allowed where it was needed. Two months later, in Budapest, the Warsaw Pact group met and refined these proposals; they further proposed that the initial reductions be 150,000 personnel and that these reductions be followed with others to a total of a half-million on each side. This would mean a reduction of a million troops at the end of 5 years.

NATO was given these proposals 10 months ago and is still working on them. How the work is going, I do not know. If there is a future for this particular kind of arms control, I think that it cannot depend entirely on manpower counts because of differences over counting rules.

Perhaps a more promising path would be to use divisions and to work out rules by which a division can be given a certain weight, taking into account all the factors that give it its strength or lack of strength (not only manpower but equipment, location, readiness, fraction of slots filled, and so forth.)

Another possibility, although perhaps not as attractive, would be to admit at the outset that each country views its own land forces in such a way that there will never be any common feeling developed between them, and, therefore, all that could be done would be that one side could propose a specific set of reductions and ask the other side what it would give for it. One cannot be either optimistic or pessimistic about this possibility, although there is a growing accumulation of background work and many of the problems are well defined. One could imagine that these efforts would proceed and that the NATO response to the Budapest initiative will start movement on that path.

Finally, there is the alternative approach to arms control in the conventional field, the operational approach, which means regulating military activities. To the surprise of many people, this alternative bore its first fruits in Stockholm last summer, where, after 3 years, and again with the intercession of Gorbachev and his sending his chief of military staff there to carry out the negotiations, there was a set of agreements reached that are extremely promising.

These agreements provide for prior notification of many important military activities; also, observations of exercises that involve more than 17,000 troops will be required. There will be an annual calendar provided by a certain date that will contain all of the maneuvers planned for the next 2 years involving more than 40,000 troops. Three onsite inspections will be allowed per year. Each of these agreements is modified and limited in certain ways that one may not like, but it is still a substantial improvement.

Attempts to extend these proposals are currently going on in Vienna and they are expected to continue for another 6 months. Here, again, however, one might find ways of limiting military activities so as to reduce the possibility of a rapid mobilization, a surprise attack, and ultimately to thin out the troop concentration around the inter-German border by moving troops back from it.

What conclusions should be drawn from all of this? To put it very briefly, there is a possibility of substantial conventional force arms control and troop reductions, but it is probably going to be a fairly long process. It may be pushed by the fact that some countries have demographic problems that will make the value of young men working in the factories and elsewhere much more important than being unproductive in military units. The possibility that these negotiations will bear fruit, I think, is much more likely now than it has been in the past.

It remains, then, to discuss how possible reductions would relate to things that are in the air today. I think my main point is clear: the uncertainty of evaluating the military effectiveness of conventional forces is so great that one cannot imagine that the elimination of intermediate-range missiles in Europe is going to call for any substantial change in the conventional forces beyond quickening the pace of the improvements now under way.

With the 50 percent reduction in strategic forces, I think one comes to the same conclusion because, although 50 percent seems like half, it really is not because the effects of using nuclear weapons are not a linear function of the number. One reaches overkill sooner or later—fairly soon—so that the last 50 percent of weapons could never have the military effect that the first 50 percent used would have.

It is for that reason that things do not change a great deal, even at 50 percent reductions. Later on, they would, but we are not, in either of these cases, able to go back to the situation that existed in the early 1950s, when we thought of nuclear weapons as a substitute for manpower.

The presence and function of manpower in conventional forces in Europe exists in its own right and is a partner with nuclear weapons in the deterrent; there is very little trade-off between nuclear weapons and troop levels possible at the levels of nuclear reductions in Europe we are now talking about.

All of this contradicts what many experts are saying recently. Representative Les Aspin, Senator Sam Nunn, former National Security Advisor Brent Scowcroft, former National Security Advisor Henry Kissinger, and former President Richard Nixon have all come out opposing in some degree or other the agreement to eliminate to zero the intermediate-range missiles—1,650 on the Soviet side and 316 on our side—a deal that one could hardly have dreamt would be possible before. Yet these individuals oppose it, and their opposition stems from two sources.

One is that some are sufficiently unhappy about the state of the conventional balance that they want to use this opportunity to improve it. In most cases, they suggest either a decrease in the number of Soviet divisions or an increase in the number of Western divisions. My argument has been, however, that the military efficiency of those forces is so uncertain and depends on so many other factors that to solve it in terms of divisions is not very satisfactory.

Therefore, I must disagree with their conclusions. Instead, I believe we should grasp the new opportunities that seem to lie before us, the beginnings of a new route in conventional arms control. Yet, the complexity of the issues involved are so substantial that one can probably expect only slow progress. But even with some progress in structural and operational arms control, conventional forces will remain largely invariant with respect to the state of nuclear weaponry and changes therein because the level of nuclear reductions that are in prospect today do not really affect the separate and important mission of the conventional forces in Europe.

7

The Future of Arms Control

Marvin L. Goldberger

Predicting the future is a mug's game. Who could have predicted a Gorbachev, a Reykjavik, or the Reagan Star Wars speech 5 years ago? But despite the ever-present possibility of the unexpected, it is important for us to think very hard about what the future might hold for arms control and the role it may play in the continuing quest for world peace.

We normally think about arms control as the effort to achieve the following objectives: (1) to reduce the risk of war; (2) to limit damage, should it occur; and (3) to lower the costs of maintaining a military establishment. Most often, we couple the term arms control with disarmament. Yet this need not necessarily be the case in the sense that, for example, the risk of war might be reduced by the deployment of a more nearly invulnerable weapon system. Without getting into Talmudic arguments, it seems clear that what we are interested in discussing is the various steps that can be taken to avoid annihilation and the destruction of our civilization. A mathematically rigorous definition of arms control is not really very interesting in this light.

The Past of Arms Control—Triumphs and Missed Opportunities

I am not an historian, and I shall make no attempt to go back to ancient times in exploring the past of arms control. Instead, I will begin with July 17, 1945, the day after the first nuclear bomb test at Almagordo. Leo Szilard sent a petition to President Truman on that day that said

> If after this war a situation is allowed to develop in the world which permits rival powers to be in uncontrolled possession of these new means of destruction, the cities of other nations will be in continuous danger of sudden annihilation. All the resources of the United States, moral and material, may have to be mobilized to prevent the advent of such a world situation.

Wise words; but unfortunately, we were not able to mobilize those moral forces, and the threat of mutual destruction is what the United States and the Soviet Union face and in fact have faced for the past 35 years or more. I shall use most of this space to discuss those aspects of arms control that relate to the special dangers to humanity posed by nuclear weapons. Yet this is only part of the problem of international security, and if we ever achieve some of the drastic cuts in nuclear weaponry that have been discussed here, many deeper issues that may be even less tractable will have to be considered.

The truly grand opportunity to virtually eliminate nuclear weapons came with the so-called Baruch plan developed by Bernard Baruch, Robert Oppenheimer, Dean Acheson, and David Lilienthal and presented to the United Nations in 1946. The proposal would have placed all the nuclear resources of the world under the ownership or control of an independent international authority, which would conduct its own inspections and would have jurisdiction over all stages of nuclear weapons manufacture. Once the machinery was in place, the United States would surrender its nuclear arsenal. The large majority of UN members supported the plan, but the Soviet Union objected to the ownership, staging, and enforcement provisions. Their own verification provisions were deemed altogether inadequate, and the negotiations became deadlocked. This was a tragedy.

The next great opportunity for a profound step was in connection with the decision in October 1949 to embark on a crash program to develop thermonuclear weapons. Many are undoubtedly familiar with the history of the famous report of the General Advisory Committee to the Atomic Energy Commission chaired by Robert Oppenheimer. This committee pointed out that there was no military requirement for such weapons because good old fission bombs could be made adequately destructive and that the prospect of building bombs with essentially unlimited destructive power was intrinsically evil and immoral. There are those who say that it was this excursion into the moral arena that undercut the report. After all, the argument went, who needs moral scientists? It would seem that an effort should have been made to outlaw the production of hydrogen bombs at that point precisely because no bombs had yet been made and verification procedures at that stage would have been relatively simple. As far as I know, however, no serious effort was made to ban that bomb.

Among the most important arms control agreements reached with the Soviet Union and the first major treaty negotiated in the nuclear area was the Limited Test Ban Treaty (LTB) of 1963. It was a triumph—and at the same time, a tragedy. The triumph was that testing of nuclear weapons in the atmosphere, in the oceans, and in outer space was prohibited. Public pressure from concern over radioactive fallout from atmospheric tests and to some extent from underwater explosions played a role in our country's seeking that accord. The verification of violations in all three media was not too difficult, although there were those who feared the Russians might cheat by testing behind the moon or the planets—a truly ridiculous notion.

The tragedy was the failure to achieve a comprehensive test ban (CTB) that would have foreclosed the option of testing nuclear weapons underground. I have talked to three people who were intimately associated with this treaty: Carl Kaysen, Spurgeon Keeny, and Bob McNamara. The reasons why the United States did not push for such a treaty despite the fact that President Kennedy wanted it are complex. We were apparently prepared to talk about a CTB when the U.S. group went to Moscow. Frank Press was brought along as an expert in connection with the thorny issue of the number of onsite inspections. The United States wanted seven or eight per year, the Soviets two or three. We might have compromised on four or five but Khrushchev did not want to discuss the matter at that time and Ormsby Gore, Kuznetsov, and Bill Foster took very rigid positions on the inspections. Kennedy was anxious to get an agreement (as was Khrushchev) after the Cuban Missile Crisis and was unsure of his political clout in being able in this country to face down Senator Richard Russell and the Joint Chiefs of Staff who were very leery about any treaty. Ultimately, with McNamara playing a central role, the Chiefs reluctantly went along with the LTB although they exacted a heavy price in dollars for maintaining a full test regime readiness (in case the treaty had to be abrogated) and a commitment to a vigorous underground test program. McNamara feels now that there was virtually no possibility of convincing them to embrace a CTB at that time. Needless to say the structure of our strategic forces would have been quite different had testing stopped in 1963; in my opinion, we may have had more stable and survivable weapons systems.

One of the early questions that came up in connection with underground testing had to do with the idea of decoupling the explosion of a nuclear weapon from the surrounding earth by setting it off in a big hole and hiding the otherwise distinct seismic signal. This technical possibility was seized on by treaty opponents who argued that the Soviets would cheat and thus gain some advantage over us. I mention this only because potential Soviet cheating is frequently raised in connection with treaties. It turns out that digging an adequate hole with dirt volumes of the order of the great pyramids is hard to do clandestinely. Another related cheating mode was to set off a bomb during one of the big earthquakes that the Soviet Union has in good supply. The point I want to make here is the absurdity of the idea, analogous to that of testing behind the moon, that by a few tests that had a high probability of being detected anyway, the Soviets could gain a serious military advantage. This issue comes up over and over again in arms control history. We always insist that we must have some large number of tests for one or another weapons development that the Soviets, we say, can accomplish with a few clandestine tests under the most difficult conditions.

There are two other treaties that have been successfully negotiated and are of the greatest importance, both historically and right now. These are the Non-Proliferation Treaty (NPT) signed in July 1968 and entered into force in March 1970, and the ABM treaty, signed in May 1972 and entered into force in October 1972.

The purpose of the NPT was to prevent the spread of nuclear weapons by forbidding the transfer of nuclear weapons to the national control of any country that did not already have them and, further, to initiate provisions designed to prevent the diversion of nuclear materials from peaceful to weapons use. I think the fact that there are today only 6 admitted possessors of nuclear weapons rather than the 20 or so one might have predicted 25 years ago is a measure of the value of the NPT. It is worth pointing out that Article VI of the treaty states that the parties will pursue in good faith measures relating to the cessation of the arms race and to nuclear disarmament. Of course, as is often noted by nonmember states, the number of weapons in the world has increased since the NPT came into force. Most people believe that a CTB prohibiting testing is the only way to prevent the emergence of additional nuclear-weapon states. There are also those who say that Israel might provide a counterexample to this belief. It might be well to remember that the weapon that destroyed Hiroshima had never been tested.

The ABM treaty of 1972 has served the true national security interests of the United States very well. There are those who disagree with this statement, but they are in the minority—as well as being wrong. Although, unfortunately, there has been a great increase in the number of strategic warheads since 1972, a full-blown offense-defense arms race has been avoided. We wasted only $7 billion on our now mothballed ABM installation at Grand Forks, North Dakota. The Soviets continue to upgrade their 100 launcher system around Moscow, but that does not really cause much concern. The treaty has allowed a research program that was actually initiated at the same time as ballistic missiles were introduced (contrary to some opinions, defense did not begin with the Reagan speech on March 23, 1983) and that has enabled us to reassure ourselves on the technical assumptions about defense capabilities that underlie the treaty. Although we might have expected that with no significant ballistic missile defense, both the United States and the Soviet Union might have cut back or at least not increased their offensive force, this has, of course, not happened. The absence of defense has also elevated the importance of and concern by the Soviet Union for the strategic nuclear forces of Great Britain, France, and China.

Let me close this review of past arms control efforts with two brief remarks. The unratified SALT II treaty of 1979 would have put a useful cap on the strategic arms race, and although it did not have the deep implications of the ABM treaty, it was important and should have been enacted. As many know, the Soviet invasion of Afghanistan caused President Carter to ask to have the treaty withdrawn from consideration for ratification in the Senate; it was later deemed by Ronald Reagan in the 1980 presidential campaign to be fatally flawed. The other comment is about the treaty that never was—the one banning MIRVs—or multiple independently targeted reentry vehicles. In 1968 much work had been done on the idea of not deploying this rapidly evolving technology which appeared to be the exclusive province of the United States. Unfortunately, before it could be explored in any serious diplomatic way, the Soviet Union invaded Czechoslovakia. This was followed by the change of administrations. Ambassador Gerry Smith actually pursued internally a so-called SWWA—Stop Where We Are—proposal that would have precluded U.S. MIRV deployment and forbidden Soviet testing and deployment. He raised the question with Secretary of State Rogers as to whether or not MIRV deployment by both sides might be a very dangerous development. Unfortunately, concern over Soviet ABM activity and the pressure of the technological imperative (whatever can be built must be built) led us to deploy MIRVs, and in the characteristically mindless manner of the arms race, the Soviets responded with their own MIRVs to penetrate what by that time was a nonexistent U.S. ABM system following the ABM treaty. Clearly, Ambassador Smith's worst fears have been borne out.

The reason for my dwelling so long on the past is that in contemplating the future of arms control it is important to recognize some of the irrationalities and peculiar forces that have shaped the present dilemma, which casts such a long shadow on the future.

The Future of Arms Control

There are many critics of arms control objectives and the arms control process. We certainly have not seen great progress in lowering the risk of war, although some helpful technical steps like the Washington-Moscow Hotline have been taken. Clearly, there has not been much in the way of limiting the damage that would result if nuclear war occurred. The latter failure is obviously connected with the awesome destructive power of nuclear weapons. As Dr. Panofsky noted in his introduction, one Trident 2 submarine with optimal targeting can cause 30 million–50 million deaths, and we might have 10 or more on patrol at any one time.

Why, then, do we persist in arms control? What kinds of alternative paths can we imagine to try to achieve the ultimate goal of removing war or the threat of it as an instrument for gaining national objectives? We could work toward changes in the international order. We could simply try to win the arms race, that is, become so powerful that the Soviets would quit racing. We could try taking unilateral steps toward disarmament and hope that the Soviets would follow suit. We could try to work toward President Reagan's impenetrable shield—put our faith in that technological "fix." The bottom line in my opinion, however, is that there is simply no way other than arms control that offers real hope for progress in the near future. On the other hand, this does not mean we should not raise questions about the process and look for new approaches, learning from the experiences of the past.

Some of the benefits of arms control are not emphasized sufficiently because they are difficult to quantify. The very negotiating process itself builds up a momentum and the hope that a current, modest-looking agreement may pave the way subsequently for a more profound one. The negotiators on the two sides and all of those who back them up—the intelligence agencies, the diplomatic establishments, and, in particular, the military leaders—eventually develop a deep commitment to what is perceived as a mutually advantageous agreement, and they become powerful advocates for the agreement as well as guardians against any proposed violations. Negotiations have the effect of reducing uncertainties about each other's forces because these must be precisely stipulated and agreed upon from the beginning. There is some reason to believe that success in the arms control arena will lead to other areas of agreement between the United States and the Soviet Union—the reverse of the kind of linkage usually talked about. For example, in April 1987 in Moscow, although the INF issue was central, Secretary of State Schultz and Foreign Minister Shevardnadze initialed an agreement on space cooperation.

Now, what of the critics who, while admitting some of the concrete achievements of arms control and the spin-offs just referred to, remain unconvinced that the positives outweigh the negatives? One of the shrillest criticisms by the hawks in the United States is that arms control treaties lull the country into such complacency that we will no longer support programs essential for our defense. This is a bizarre argument that is insulting to the military and political leaders entrusted with our national security. The facts of the matter simply do not support these worries. The military in the past has ensured a commitment to undiminished or even increased spending on relevant programs. The so-called safeguard provisions I noted in connection with the limited test ban treaty are a good example.

The dovish critics of arms control make a different argument. They accept the premise that one must have something to bargain with if the negotiations are to be carried out on the assumption that additional weapons or superior performance would be advantageous to the possessor. The problem the doves see is that the need for bargaining chips may lead one side or the other or both to acquire unneeded weapons. This aspect of the process is exacerbated by the inordinate length, at least historically, of negotiations, during which weapons systems acquire momentum and constituencies and become virtually unstoppable. Finally, weapons that are not included in a particular agreement are pursued with unusual vigor to combat potential critics. Some critics thus view the whole arms control process as a means for legitimizing the arms race.

The slow pace of the traditional arms control process has occasionally led some people to suggest that the United States try to be somewhat bolder in unilateral disarmament initiatives, one of the alternative paths I mentioned earlier. One can argue that the size and relative survivability of our strategic forces is so great that even if the Soviets were to continue building and we were to stop, it would make no significant difference in the strategic confrontation, particularly today, when there are no defensive systems. There are naturally serious questions about such a proposal. How great a force disparity could we tolerate without incurring military risk? Would anything like this be politically acceptable?

There is another type of unilateral action we could take that is related to the discipline we exercise on the evolution of new weapons systems and the seriousness with which we take the arms control implications of these systems. Let me list some questions that should always be addressed in this connection. I have a feeling I first heard of this approach from a CISAC member—Dr. Garwin, perhaps, or Dr. Panofsky. Nevertheless, with all due credit to whomever, the questions are pretty obvious.

- If a new system were developed by the United States and a few years later was initiated by the Soviets, would our net security be served by having it deployed against us? (MIRVs are an example of a system that might have failed this test.)
- Would a new system (for example, a hard-site ABM system with a break-out potential) make existing or projected arms control agreements more difficult to police?
- Would a system being proposed as a bargaining chip (e.g., Grand Forks, North Dakota) be one we really wanted on its merits?
- How does a proposed new system affect strategic stability? Will it increase or decrease the advantage of striking first? If it enhances the advantage to a first striker, how does it help our security?
- Does the system have obvious countermeasures? Is it cheaper to counter than to build? (Need I mention SDI in this context as one that might flunk this test?)

Richard Garwin has remarked that one must make such considerations a part of the educational process for our future military leaders. In addition, there should not be an adversarial relationship between the military and the Arms Control and Disarmament Agency or any other part of the national security establishment.

In suggesting the need for better criteria for weapons, I am reminded of some of the arguments supporting the MX: (1) it was necessary for us to proceed with the MX to demonstrate national will; (2) it was necessary to match the Soviet hard-site kill capability; and (3) it was necessary as a bargaining chip. These arguments struck me as being, respectively, silly, irrational, and cynical, hardly the prescription for sound decision making.

Let me turn now to some things that seem likely to be on the arms control agenda in the future and then conclude with some general observations on several topics that are not usually subsumed under the arms control rubric but that are, in fact, terribly important.

First, there is the series of issues raised by Reykjavik, issues that are largely unresolved as yet. On the basis of the Shultz visit to Moscow and subsequent events, there is a reasonable presumption that an agreement is in the offing on intermediate-range nuclear forces (INF) and shorter-range missiles. Obviously, in the words of Yogi Berra, it's not over 'til it's over, and it may yet be déjà vu all over again. Militarily, the weapons systems involved are in my view of dubious value, but the negotiating movement such an agreement would produce would be important. The symbolic significance of literally destroying a weapons system would be quite significant.

Clearly, as evidenced by this seminar, the concept of drastic cuts in strategic nuclear forces that was touted in various although not always clear forms at Reykjavik is very much at the center of current interest. Given the complexities alluded to in our discussions here, it is clear that this item will be with us for quite a while. I am somewhat amused to recall in connection with deep cuts that there were many people who hooted in some derision when George Kennan, about 6 years ago, suggested that we should cut strategic forces by half.

The recent Soviet proposal that the United States and the Soviet Union conduct weapons tests at each other's test site is very interesting. If such tests come to pass, the way may be paved, by increasing verification confidence, first for ratification by the United States of the Threshold Test Ban and the Peaceful Uses treaties and then perhaps serious motion toward a CTB. Given the relative maturity of nuclear weapons design and the possibility of assessing weapons reliability without tests, it is hard to believe that our security or that of the Soviets would be threatened by a CTB. The gains are significant psychologically for the rest of the world and the non-NPT nations in particular. In addition, of course, various potentially mischievous weapons developments—like x-ray lasers, for example—could be foreclosed.

Obviously, the whole set of issues associated with strategic defense will be part of any future arms control agenda. For the moment the critical item is preserving the integrity of the ABM treaty. The so-called broad or legally correct interpretation of the treaty advocated in some quarters of the Reagan Administration would appear to be a major obstacle to achieving any of the significant cuts in offensive forces discussed here. The role of defense in an era of reduced offenses will require much analysis and eventually difficult negotiations.

There are a number of other obvious foci for arms control activity in the years to come, including, for example, antitactical missile defenses, cruise missile issues, the implications of mobile or concealable strategic weapons, conventional force reductions, and multilateral negotiations involving all the major nuclear powers, particularly if the superpowers significantly cut their strategic forces. There is more than enough work to be done, which leads me to the final portion of my discussion.

The current unsatisfactory, unstable, and dangerous situation in which we find ourselves did not develop suddenly. In both the United States and the Soviet Union the situation evolved through a series of steps that frequently were taken for domestic or international political reasons virtually unconnected to real strategic military objectives. There are a number of culprits: the scientists who failed to communicate adequately to the public the realities of nuclear weaponry; the infatuation of the military with the power of these new gadgets and their inability to recognize that there was no way to use this power—to capture a city or a country destroyed in a nuclear war is not really very important; the dynamics and momentum of the arms race in the United States; the military-congressional-industrial complex; the xenophobia and insecurity of the Soviets; their past and present harsh and aggressive policies, both internally and toward their neighbors; their own military-industrial complex; and the general aura of mistrust with which the United States and the Soviet Union regard each other. Altogether, there are no shortages of explanations; nevertheless, it is often hard to accept the collective madness that has led us to where we are.

The arms control efforts we have engaged in for nearly 30 years have been aimed at extricating us from this mess in a safe and stable way. In spite of the immense amount of work that has gone into the effort, I feel that what has been lacking is a coherent, clear-cut, and overarching long-range vision of what our objectives are. We have attempted to do things one at a time—first this kind of a treaty, then that one, and so forth. I once tried to pose the following question to CISAC and, I think, also to our Soviet counterparts: The time is *n* years from now, and the United States and the Soviet Union are at peace in a crisis-stable and secure relationship. Can we define that new era in some detail, that is, what is *n*, and how do we get there from here? The question is easy to pose but not as easy to address. In this country, we have a recurring problem in long-range planning caused by the presidential electoral process. Thus, we practice arms control interruptus every 4 years. Institutions outside the government such as the universities and the various think tanks can help develop long-range policy. But there is a strong resistance on the part of people in government to outside advice, and a mechanism for analogous considerations is also necessary inside the system.

In the short run, we have to survive; thus, the kind of technical arms control efforts I discussed earlier must be pursued vigorously. In a real sense, these things, although very important, are secondary issues. We are actually only buying time. There are no technical solutions to our dilemma; there really are no impenetrable shields or magic bullets. World peace will have to be based on many factors that lie outside traditional arms control. Things like technological change, regional rivalries, Third World sociopolitical evolution, and the diminishing oil supply are all part of the picture, which is a rapidly changing one. Technical and economic forces are dramatically changing our biosphere; witness the destruction of the rain forests, the destruction of the ozone layer, and the CO_2 buildup. All of these issues in various ways will have a profound influence on international security, more profound in the long run than that posed by nuclear weapons. The effects of history, the politics and sociology of strategic rivalries, the linkages between economics and defense, and the stresses posed by the inequalities of living standards are other aspects of the problem. Tackling these issues is going to require the combined efforts of physical and life scientists as well as political and social scientists, engineers, lawyers, doctors, and businessmen. The phenomena are truly complex and the time available for understanding them and translating this understanding into policies to avoid disaster is very short. If war doesn't get us, overpopulation or some form of environmental disaster will. Instead of trying to make nuclear weapons impotent and obsolete, let's work to make them irrelevant.

We scientists have a special responsibility. We must help to develop in a public that is largely scientifically and technically illiterate an understanding of nuclear weapon realities and of the host of major questions with a strong technical component. In our universities, we must help to educate students with a serious and broad background in security issues. There is more than enough for us to do.

The problem of trying to achieve a peaceful world has been with us for a long time. If we do not blow ourselves up, it will continue for a long time. But it is important to remember that we are not dealing with a physics problem that we can abandon if it gets too tough and then work on something else. This is the problem; it will not go away, and we cannot allow ourselves to become discouraged, easy though that may be. There are some glimmerings of hope in the air and the absurdity of a presumably civilized society wasting nearly a trillion dollars a year on arms throughout the world, to say nothing of the human talent diverted from the real problems of survival on an underresourced and overpopulated planet, finally seems to be sinking in. As Victor Hugo said, "An invasion of armies can be resisted, but not an idea whose time has come."

Let's get on with it!